Electrons in Metals and Semiconductors

PHYSICS AND ITS APPLICATIONS

Series Editor

E.R. Dobbs
University of London

This series of short texts on advanced topics for students, scientists and engineers will appeal to readers seeking to broaden their knowledge of the physics underlying modern technology.

Each text provides a concise review of the fundamental physics and current developments in the area, with references to treatises and the primary literature to facilitate further study. Additionally texts providing a core course in physics are included to form a ready reference collection.

The rapid pace of technological change today is based on the most recent scientific advances. This series is, therefore, particularly suitable for those engaged in research and development, who frequently require a rapid summary of another topic in physics or a new application of physical principles in their work. Many of the texts will also be suitable for final year undergraduate and postgraduate courses.

Electrons in Metals and Semiconductors

R.G. Chambers
Professor of Physics
University of Bristol

CHAPMAN AND HALL
LONDON • NEW YORK • TOKYO • MELBOURNE • MADRAS

UK	Chapman and Hall, 11 New Fetter Lane, London EC4P 4EE
USA	Chapman and Hall, 29 West 35th Street, New York 10001
JAPAN	Chapman and Hall Japan, Thomson Publishing Japan, Hirakawacho Nemoto Building, 7F, 1-7-11 Hirakawa-cho, Chiyoda-ku, Tokyo 102
AUSTRALIA	Chapman and Hall Australia, Thomas Nelson Australia 480 La Trobe Street, PO Box 4725, Melbourne 3000
INDIA	Chapman and Hall India, R. Sheshadri, 32 Second Main Road, CIT East, Madras 600 035

Typeset in 10/12 Times by
Thomson Press (India) Ltd, New Delhi
Printed in Great Britain by
St Edmundsbury Press, Bury St Edmunds, Suffolk.

ISBN 0 412 36840 4

British Library Cataloguing in Publication Data

Chambers, R. G.
Electrons in metals and semiconductors.
1. Semiconductors, electrons.
I. Title
530.41

ISBN 0 412 36840 4

Library of Congress Cataloging in Publication Data Available

Contents

Preface

Solid-state physics has for many years been one of the largest and most active areas of research in physics, and the physics of metals and semiconductors has in turn been one of the largest and most active areas in solid-state physics. Despite this, it is an area in which new and quite unexpected phenomena – such as the quantum Hall effect – are still being discovered, and in which many things are not yet fully understood. It forms an essential part of any undergraduate physics course.

A number of textbooks on solid-state physics have appeared over the years and, because the subject has now grown so large, the books too have usually been large. By aiming at a more limited range of topics, I have tried in this book to cover them within a reasonably small compass. But I have also tried to avoid the phrase 'It can be shown that...', as far as possible, and instead to explain to the reader just why things are the way they are; and sometimes this takes a little longer. I hope that some readers at least will find this approach helpful.

1

The free-electron model

1.1 THE CLASSICAL DRUDE THEORY

The characteristic properties of metals and semiconductors are due to their conduction electrons: the electrons in the outermost atomic shells, which in the solid state are no longer bound to individual atoms, but are free to wander through the solid. A proper understanding of metallic or semiconducting behaviour could not begin, obviously, until the electron had been discovered by J.J. Thomson in 1897, but once this had happened, the significance of the discovery was at once recognized. By 1900 Drude had already produced an electron theory of electrical and thermal conduction in metals, which (with refinements by Lorentz a few years later) survived until 1928. Not surprisingly, this very early theory did not manage to explain everything – after all, the structure of the atom itself was quite unknown until Rutherford and his co-workers discovered the nucleus in 1911 – but it did have one or two striking successes, and it is worth starting with a brief look at this classical model, because it already contained many of the right ideas.

In the Drude model, each atom is assumed to contribute one electron (or possibly more than one) to the 'gas' of mobile conduction electrons. The remaining positive ions form a crystal lattice, through which the conduction electrons can move more-or-less freely. This gas of conduction electrons differs from an ordinary gas (e.g. O_2) in three main ways. First, the gas particles – the electrons – are far lighter than an ordinary gas molecule. Secondly, they carry an electric charge. Thirdly, they are travelling through the lattice of positive ions, rather than through empty space, and presumably are colliding constantly with these positive ions. They may also collide with each other, as ordinary gas molecules do, but we can expect these electron–electron collisions to be less frequent, and less important, than the electron–ion collisions.

We can work out the properties of this model very easily, using the kinetic theory of gases. If m and $\boldsymbol{v}$ are the mass and velocity of an electron, then according to classical physics the average kinetic energy at temperature T is given by

$$\tfrac{1}{2}mv_r^2 = \tfrac{3}{2}kT \tag{1.1}$$

where k is Boltzmann's constant and v_r^2 denotes the average value of $|\boldsymbol{v}|^2$ over all the conduction electrons, so that v_r is their rms speed. (Note that (1.1) is in fact *wrong* for electrons in a metal: as we shall see later, quantum mechanics gives a different answer.) Every so often the electrons will collide with the ions of the crystal lattice. We assume that between collisions an electron travels with constant velocity $\boldsymbol{v}$, and that the effect of a collision is to randomize the direction of $\boldsymbol{v}$, but to leave its magnitude v practically unchanged, because the ions are far heavier than the electrons. For any one electron, the collisions occur at more or less random intervals, and the average time interval between collisions is called the relaxation time, τ. The corresponding average distance between collisions, $l = \tau v$, is called the mean free path.

If we take a snapshot of the electron gas at a given instant, and focus on a particular group of electrons all having the same speed v, some of them will have just collided, others will be just about to collide, and most will be somewhere in between. On the average, in fact, they will be just half-way between collisions, so we might expect them, on the average, to have travelled for a time $\frac{1}{2}\tau$ and a distance $\frac{1}{2}l$ since they last collided. But this is wrong: in our snapshot, we are much more likely to catch any given electron in a long free path than in a short one, precisely because it spends much more of its time traversing long free paths than short ones. This means, as we shall show in section 8.3, that the electrons in our snapshot have on average travelled for a time τ and a distance $l = \tau v$ since they last collided – twice as far as we expected.

Using these ideas, we can now work out the electrical conductivity σ of a metal on the Drude model; that is, we can work out the current density $\boldsymbol{J} = \sigma \boldsymbol{E}$ produced by an electric field $\boldsymbol{E}$. Macroscopically, $\boldsymbol{J}$ is given by $\boldsymbol{J} = \rho_c \boldsymbol{v}_d$, where $\boldsymbol{v}_d$ is the velocity at which the charge density ρ_c is moving. If our metal contains n

conduction electrons per unit volume, each carrying charge $-e$, we have $\rho_c = -ne$ and $\boldsymbol{J} = -ne\boldsymbol{v}_d$ if all the electrons have the same velocity. More generally, if the ith electron has velocity $\boldsymbol{v}_i$, we have

$$\boldsymbol{J} = -e\sum_i^n \boldsymbol{v}_i \tag{1.2}$$

If $\boldsymbol{E} = 0$, the electron gas as a whole is in thermal equilibrium, with as many electrons moving to the right as to the left, so that $\sum \boldsymbol{v}_i = 0$ and there is no net current. But in a field $\boldsymbol{E}$, each electron experiences a force $-e\boldsymbol{E}$, and an acceleration $-e\boldsymbol{E}/m$. If, at any instant, the average electron has travelled for a time τ since its last collision, it will therefore have acquired a drift velocity

$$\boldsymbol{v}_d = -e\boldsymbol{E}\tau/m \tag{1.3}$$

in addition to its thermal velocity $\boldsymbol{v}$, and we have

$$\boldsymbol{J} = -ne\boldsymbol{v}_d = (ne^2\tau/m)\boldsymbol{E} \tag{1.4}$$

so that

$$\sigma = ne^2\tau/m \tag{1.5}$$

if we assume that the relaxation time τ is the same for all electrons.

If instead, we assume that the mean free path l is the same for all electrons (and this would be more plausible on the Drude model), (1.3) becomes

$$\boldsymbol{v}_d = -e\boldsymbol{E}l/mv, \tag{1.6}$$

so that $\boldsymbol{v}_d$ varies with v, and we have to be a little careful in evaluating (1.2), if we want an exact solution (see section 2.6). But approximately, we can replace τ in (1.5) by l/v_r, since most electrons will have speeds not too different from the rms speed v_r given by (1.1), so that we have

$$\sigma \approx ne^2 l/mv_r \tag{1.7}$$

as the Drude expression for the electrical conductivity. We can not use this to calculate a theoretical value of σ, though, because the Drude theory does not predict a value for l; but what we *can* do is to use (1.7) to calculate l, and see if the result is plausible.

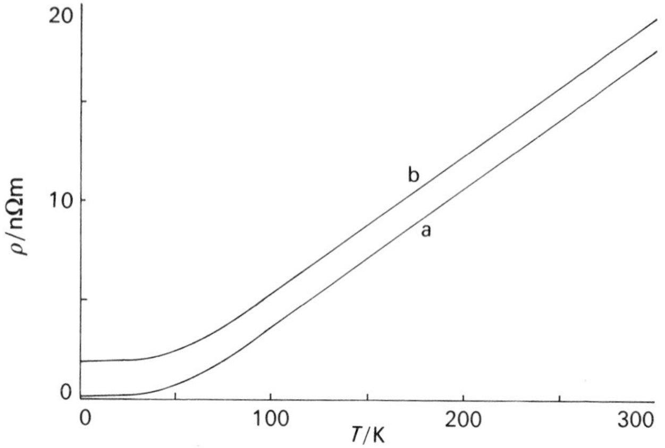

Fig. 1.1 The resistivity $\rho(T)$ of two Cu samples. The residual resistance ratio $\rho(293)/\rho(0)$ is 100 for sample a and 10 for sample b. In a pure enough sample, the r.r.r. may be over 1000.

If we do this for say Cu at room temperature (problem 1.2), assuming one conduction electron per atom, we find $l \sim 3$ nm or 30 Å. Since the separation between nearest neighbour atoms in Cu is 2.55 Å (0.255 nm), this would have seemed quite a reasonable value for l in Drude's day, though as we shall see later, l is now known to be an order of magnitude greater than this.

The temperature dependence of σ is more difficult to account for on the Drude theory. In (1.7), only v_r would be expected to depend much on temperature: from (1.1), we have $v_r \propto T^{1/2}$, so we expect $\sigma \propto T^{-1/2}$ or $\rho \propto T^{1/2}$ where $\rho = 1/\sigma$ is the electrical resistivity. But for most metals ρ varies roughly as T at temperatures above about 200 K, and much more rapidly (in pure metals) at low temperatures (Fig. 1.1). To account for this on the Drude theory, we would have to suppose that either n or l (or both) becomes much larger at low temperatures, which does not seem very plausible on the classical picture.

To calculate the thermal conductivity κ, Drude simply used the well-known result from the kinetic theory of gases,

$$\kappa = \tfrac{1}{3} C_v \bar{v} l \tag{1.8}$$

where $\bar{v}$ is the mean speed of the conduction electrons and C_v is the heat capacity of the electron gas per unit volume. Classically, we expect the mean thermal energy per electron to be given by (1.1), so that $C_v = \frac{3}{2}nk$ if there are n electrons per unit volume. We then have

$$\begin{aligned}\kappa &= \tfrac{1}{2}nk\bar{v}l \\ &\approx \tfrac{1}{2}nkv_r l \qquad (1.9)\end{aligned}$$

since $\bar{v}$ and v_r differ by only a few percent.

As before, we can use (1.9) to work out l from the measured value of κ. For Cu at room temperature, we find $l \sim 6$ nm or 60 Å – rather bigger than the value from σ, but still not too unreasonable. But as before, the temperature dependence of κ is given less well; (1.9) suggests that κ (like v_r) should increase as $T^{1/2}$, whereas experimentally κ behaves in a considerably more complex fashion (Fig. 1.2).

The theory has more success in predicting the ratio κ/σ or $\kappa/\sigma T$. From (1.7) and (1.9) we have

$$\begin{aligned}\kappa/\sigma T &= mv_r^2 k/2e^2 T \\ &= 3k^2/2e^2, \qquad (1.10)\end{aligned}$$

using (1.1). The theory thus predicts that the ratio $\kappa/\sigma T$ should be

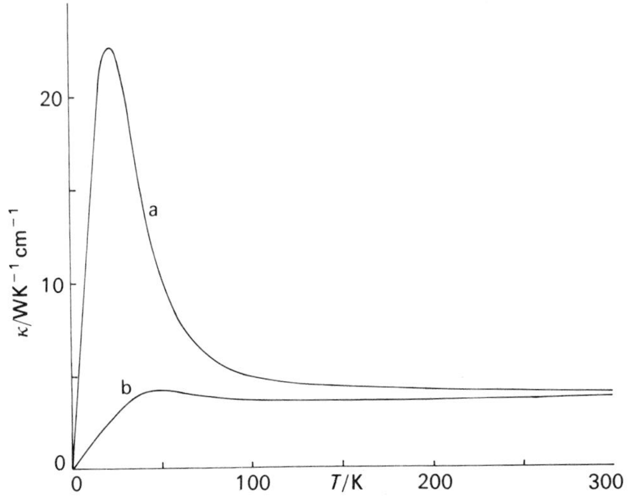

Fig. 1.2 The thermal conductivity $\kappa(T)$ of the same two Cu samples as in Fig. 1.1.

a universal constant, the same for all metals. This was a major triumph of the theory, because exactly this result had been observed experimentally by Wiedemann and Franz in 1853 and by Lorenz in 1872; the ratio $\kappa/\sigma T$ is called the Lorenz number $\mathscr{L}$.

As it happens, Drude's predicted value of $\mathscr{L}$ was not $3k^2/2e^2$ but $3k^2/e^2$, because in calculating $\boldsymbol{v}_d$ (equation 1.3) he had mistakenly assumed a drift time of $\frac{1}{2}\tau$ instead of τ. But as it happens, $3k^2/e^2$ ($= 2.23 \times 10^{-8}\,V^2\,K^{-2}$) is much closer to the experimental values, which lie between 2.2 and $2.6 \times 10^{-8}\,V^2\,K^{-2}$ for most metals at and above room temperature. So, by a happy accident Drude's theory appeared to be in remarkable agreement with experiment. When Lorentz (not to be confused with Lorenz) worked out the classical theory exactly in 1905, avoiding the numerous approximations that Drude had made (and that we have made), he found $\mathscr{L} = 2k^2/e^2$; disappointingly, not such good agreement with experiment.

Moreover, the theory predicts that $\mathscr{L}$ should be independent of temperature, whereas this is not so experimentally, as Fig. 1.3 shows. $\mathscr{L}$ is reasonably constant at high temperatures, but then falls, to

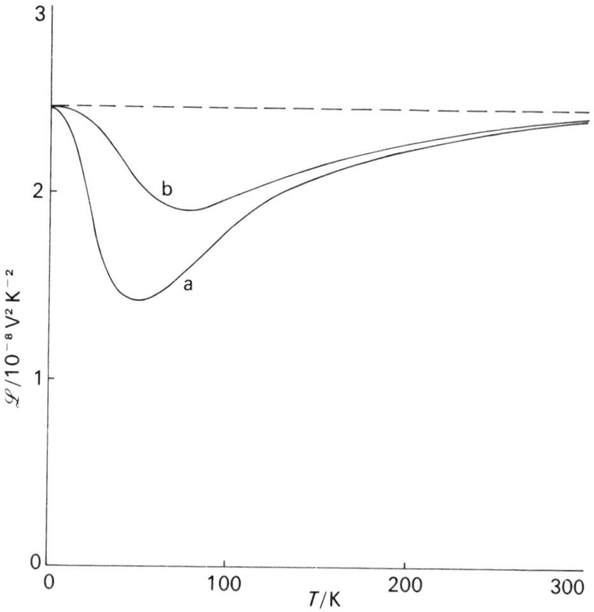

Fig. 1.3 The Lorenz number $\mathscr{L}(T)$ for the same two Cu samples as in Figs 1.1, 1.2.

return again at very low temperatures to something like the high-temperature value. Nevertheless, the theoretical predictions were close enough to the observed facts to suggest that the Drude theory must contain a good deal of the truth, if not the whole truth. But there was at least one gross discrepancy between the Drude theory and the observed facts, and that concerned the heat capacity of metals. In deriving (1.9), we assumed that each electron in the conduction electron gas contributed an amount $3k/2$ to the heat capacity. Now in classical physics, the thermal vibration of each atom in a solid (or each ion, in a metal) must contribute $3k$ to the heat capacity, so that the heat capacity per mole of a monatomic solid should be $3N_A k = 3R$ (where N_A is Avogadro's number), in agreement with Dulong and Petit's law. But if a mole of metal also contains about N_A conduction electrons, as Drude assumed, then classically they should contribute an additional $3R/2$ to the heat capacity: the molar heat capacity of metals should be about half as big again as that of insulating solids. Experimentally, there was no sign whatever of this additional contribution to the heat capacity. This discrepancy was quite inexplicable in classical terms, and further progress had to await the advent of quantum mechanics.

1.2 FERMI–DIRAC STATISTICS

By 1926 it was known that electrons were spin-$\frac{1}{2}$ particles obeying the laws of quantum mechanics and, in particular, obeying the Pauli exclusion principle. In 1926 Fermi and Dirac independently pointed out that a perfect gas of electrons would consequently behave very differently, at least at high densities, from a classical perfect gas. Because of the exclusion principle, not more than one electron can occupy any given quantum state, if we count spin-up and spin-down states as different. If we imagine electrons being poured into an initially empty box at $T = 0$, the first two will go into the two lowest-energy states (one with spin up and one with spin down), the next two will have to go into the next two higher-energy states, and so on. When we have poured in N electrons, all the N states of lowest energy will be filled, and all the higher-energy states will be empty.

If we really did this, of course, the box would acquire an enormous negative charge. To avoid this, we imagine that at the same time we pour in an amount Ne of positive charge, to keep the box neutral, and to avoid complications we assume for now that this positive charge is smeared out into a uniform continuum instead of being concentrated into positive ions.

There is yet another simplifying assumption that we need to make: in considering the motion of any one electron, we assume that all the other electrons, too, are smeared out into a uniform continuum of negative charge. In other words, we ignore all the details of the Coulomb interaction between individual electrons. We have to ignore them, because otherwise it would be impossible to talk of any individual electron being in a definite quantum state, with a definite energy.

Exactly the same problem arises in the kinetic theory of gases: in real gases, the molecules exert forces on one another, and this makes the theory of real, imperfect gases very difficult. We can simplify matters by neglecting the intermolecular forces and dealing instead with perfect gases. Here we do the same, and consequently the Fermi–Dirac theory is the theory of a *perfect gas* of non-interacting electrons.

So: we now have a box containing N non-interacting electrons, filling up the lowest N available states. If $f_0(\varepsilon)$ denotes the probability that a state of energy ε is occupied, we have $f_0 = 1$ for all states up to some particular energy $\varepsilon_{F,0}$ called the Fermi energy, and $f_0 = 0$ for all states of higher energy. This is the situation at $T = 0$, when the system as a whole has the lowest possible energy. Because of the exclusion principle, it is totally different from the situation which would exist in a classical perfect gas at $T = 0$, where all the particles would have zero energy.

What happens now to our electron gas as we warm it up? What happens is that the abrupt drop from $f_0 = 1$ to $f_0 = 0$ at $\varepsilon = \varepsilon_{F,0}$ becomes blurred out – only slightly if $kT \ll \varepsilon_{F,0}$, but increasingly as kT increases. The form of $f_0(\varepsilon)$ is given by the Fermi–Dirac function

$$f_0(\varepsilon) = \frac{1}{e^{(\varepsilon - \varepsilon_F)/kT} + 1} \tag{1.11}$$

where the Fermi energy ε_F (sometimes written μ and called the chemical potential) depends on T and on the total number of electrons N. If we sum $f_0(\varepsilon)$ over all the states of the system, we must have

$$\sum f_0(\varepsilon) = N \tag{1.12}$$

and at given temperature this condition determines ε_F.

Though (1.11) is one of the fundamental equations in this book, it would take us too far afield to derive it here; you will find a derivation in any good book on statistical mechanics, for example *Statistical Physics* by Guénault, in this series. But we can easily see that it behaves in just the way we would expect. At $\varepsilon = \varepsilon_F$, we have $f_0(\varepsilon) = 0.5$; at $\varepsilon = \varepsilon_F - 2kT$, $f_0(\varepsilon) \approx 0.9$, and at $\varepsilon = \varepsilon_F + 2kT$, $f_0(\varepsilon) \approx 0.1$. Thus the drop from $f_0(\varepsilon) \approx 1$ to $f_0(\varepsilon) \approx 0$ occurs over an energy range of a few kT around ε_F, as shown in Fig. 1.4. For $T \to 0$, $f_0(\varepsilon)$ drops discontinuously from 1 to 0 at $\varepsilon_{F,0}$ (the zero-temperature value of ε_F), as expected.

In the 'classical limit', $f_0(\varepsilon)$ also behaves as expected. What does that mean? If we had a gas of classical particles, not obeying the exclusion principle, we should expect $f_0(\varepsilon)$ to be given by the Boltzmann factor:

$$f_0(\varepsilon) = C\,\mathrm{e}^{-\varepsilon/kT} \tag{1.13}$$

where C is a constant. The F–D function differs from this for two closely related reasons: the particles are treated as totally indistinguishable from one another, and they obey the exclusion principle. But these factors become unimportant if $f_0(\varepsilon) \ll 1$ for all ε: if there is only a small chance of one particle occupying a given state, there will be an even smaller chance of two trying to do so, and the fact that they are forbidden to do so ceases to be important;

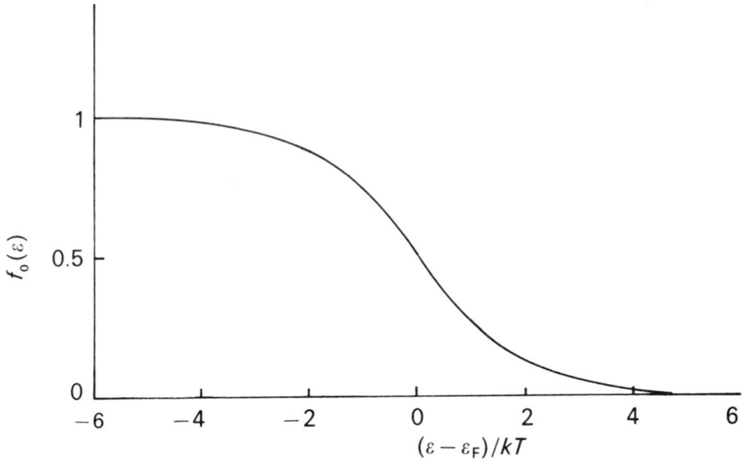

Fig. 1.4 The Fermi–Dirac function $f_0(\varepsilon)$.

and so, for most purposes, does the fact that they are indistinguishable.

Now if we are to have $f_0(\varepsilon) \ll 1$ for all ε, we need $\exp(\varepsilon - \varepsilon_F)/kT \gg 1$ for all ε, and if the lowest-energy state has $\varepsilon = 0$, this means that we need ε_F to be negative, and $|\varepsilon_F| \gg kT$. If this condition is satisfied, we can neglect the $+1$ in the denominator of (1.11), and we have

$$f_0(\varepsilon) = e^{-|\varepsilon_F|/kT} e^{-\varepsilon/kT} \tag{1.14}$$

– precisely of the form (1.13), with $C = e^{-|\varepsilon_F|/kT} \ll 1$. (As before, the value of ε_F will be determined by the condition (1.12).) Thus $f_0(\varepsilon)$ does revert to the classical Boltzmann form just where we would expect it to.

In semiconductors, the density of conduction electrons is usually so low that the electron gas is effectively in the classical limit. In metals, on the other hand, the density is so high that ε_F/kT is large and positive, and $f_0(\varepsilon)$ falls rapidly from 1 to 0 near ε_F: the gas is said to be 'degenerate'. In the next few sections, we shall be looking mainly at degenerate conductors, i.e. metals rather than semiconductors.

In the degenerate limit, there is a useful result due to Sommerfeld for evaluating integrals involving a product of $f_0(\varepsilon)$ and some function $G(\varepsilon)$. We can write

$$\int_{-\infty}^{\infty} f_0(\varepsilon)G(\varepsilon)\,d\varepsilon = \int_{-\infty}^{\infty} h(\varepsilon)G(\varepsilon)\,d\varepsilon + \int_{-\infty}^{\infty} [f_0(\varepsilon) - h(\varepsilon)]G(\varepsilon)\,d\varepsilon$$

where $h(\varepsilon) = 1$ for $\varepsilon \leqslant \varepsilon_F$ and $h(\varepsilon) = 0$ for $\varepsilon > \varepsilon_F$, so that $f_0(\varepsilon) - h(\varepsilon) \sim 0$ except close to ε_F. We have formally written the lower limit as $-\infty$, though $G(\varepsilon)$ will in practice always be zero below some finite lower limit of energy. The first term on the right is simply $\int_{-\infty}^{\varepsilon_F} G(\varepsilon)\,d\varepsilon$, and in the second term we can expand $G(\varepsilon)$ in a Taylor series about ε_F, and then integrate term by term. Since $f_0(\varepsilon) - h(\varepsilon)$ is an odd function of $\varepsilon - \varepsilon_F$, only alternate terms in the Taylor expansion survive, and it is easy to see that these will vary as $T^2, T^4, T^6 \ldots$. It is less easy to evaluate them exactly (problem 1.4), and we simply quote the result:

$$\int_{-\infty}^{\infty} f_0(\varepsilon)G(\varepsilon)\,d\varepsilon = \int_{-\infty}^{\varepsilon_F} G(\varepsilon)\,d\varepsilon + \frac{\pi^2}{6}k^2T^2\left[\frac{dG(\varepsilon)}{d\varepsilon}\right]_{\varepsilon_F} + \cdots \tag{1.15}$$

We shall never need any of the higher terms in the expansion.

1.3 THE SOMMERFELD MODEL

In 1905, Lorentz had worked out the consequences of the Drude model rather carefully, assuming that the electrons were distributed in energy according to the classical expression (1.13). In 1928, Sommerfeld in effect repeated Lorentz's calculations, but using the quantum mechanical distribution function (1.11), and thus started the quantum theory of the electronic properties of solids. He used the very simple model described in section 1.2: a box containing N non-interacting electrons, with a uniform distribution of positive charge to keep the system neutral. Each electron is thus moving in what is effectively an empty box; the effect of the other electrons, smeared into a uniform distribution, is merely to neutralize the positive charge. Obviously, this is a grossly oversimplified model, and we shall do better later, but as we shall see it is already a considerable improvement on the Drude model.

Suppose our box is a cube of side L, and that the electrons are confined within it by infinitely high potential barriers on all sides. (More realistically, we could make the barriers of finite height; this would complicate the algebra a little, without altering the results significantly.)

If we take the (uniform) potential within the box as zero, the Schrödinger equation becomes simply

$$-\frac{\hbar^2}{2m}\nabla^2\psi = \varepsilon\psi \qquad (1.16)$$

and ψ has to vanish on each face of the cube. As you probably know, the solutions to this equation are of the form

$$\psi = C\sin(\pi n_1 x/L)\sin(\pi n_2 y/L)\sin(\pi n_3 z/L) \qquad (1.17)$$

where C is a normalization constant and n_1, n_2, n_3 are positive integers, so that ψ vanishes on the faces of the cube as required. (Negative values of n_1, n_2, n_3 are excluded because they merely change the sign of ψ.) Inserting (1.17) into (1.16) we find that the energy of the state (n_1, n_2, n_3) is

$$\varepsilon = \pi^2\hbar^2(n_1^2 + n_2^2 + n_3^2)/2mL^2 \qquad (1.18)$$

But these standing-wave states are not ideal for our purpose, because we want to discuss electric currents and heat currents through the metal, and standing waves cannot carry a current. It is not surprising that these solutions cannot carry a current – they are the solutions for an isolated block of metal. We need to think again.

Suppose we make our block of metal very much longer in the x direction than in the y and z directions, so that $L_x \gg L_y, L_z$; suppose in fact that we make it a long wire, and then bend it round and join the ends together to form a ring. Then in the x direction (measured around the ring) we no longer need $\psi = 0$ at $x = 0$ and $x = L$; all we need is that $\psi(x = L) = \psi(x = 0)$, since these two ends now coincide. A suitable solution of (1.16) is now

$$\psi = C \exp i(2\pi n_1 x/L) \sin(\pi n_2 y/L) \sin(\pi n_3 z/L) \qquad (1.19)$$

and this state does correspond to an electron moving in the x direction, i.e. carrying a current around the ring. But this state is still a bit inconvenient; we want to be able to discuss currents flowing not just in the x direction, but in the y or z direction or indeed in any direction through the metal. We can do this if we go back to our cube of metal, ignore the boundary conditions on ψ for the moment, and just use running-wave solutions of the form

$$\psi = Ce^{i\boldsymbol{k}\cdot\boldsymbol{r}} \qquad (1.20)$$

This wave-function corresponds to an electron travelling in the direction of the wave-vector $\boldsymbol{k}$ with momentum $\hbar\boldsymbol{k}$ and velocity $\boldsymbol{v} = \hbar\boldsymbol{k}/m$. We certainly cannot satisfy the boundary condition $\psi = 0$ on any of the faces of the cube with this wave-function; instead, we apply to all six faces of the cube the 'periodic boundary conditions' that we imposed on the x faces in deriving (1.19). That is, we require the components of $\boldsymbol{k}$ to satisfy

$$k_x = 2\pi n_1/L, \quad k_y = 2\pi n_2/L, \quad k_z = 2\pi n_3/L \qquad (1.21)$$

where now n_1, n_2, n_3 can be positive or negative integers, so that

$$\psi = C \exp[2\pi i(n_1 x + n_2 y + n_3 z)/L] \qquad (1.22)$$

This function has the same value at any point on one face of the cube as at the corresponding point on the opposite face, so that (if it were physically possible to do so) any two opposite faces could be joined together to form a current-carrying ring. All this may sound rather unconvincing, but in fact (1.20) is a much more useful starting point for discussing the behaviour of electrons in real metals or semiconductors than (1.17). It is often useful to think of an electron as a localized wave-packet, built up from a set of $e^{i\boldsymbol{k}\cdot\boldsymbol{r}}$ states with adjacent values of $\boldsymbol{k}$, and this wave-packet will then move through the conductor, and respond to applied fields, much like a classical electron. Moreover, it will travel for only a finite distance through a real conductor – about one mean free path – before being scattered, and if the mean free path is small compared with the sample size (as it usually is) it will spend most of its time far from the surface, so that physically we would expect its behaviour to be unaffected by the surface boundary conditions.

1.4 THE DENSITY OF STATES

The importance of the surface boundary conditions is that they determine the total number of electrons that a given volume of metal can accommodate, in a given range of energies. The periodic boundary conditions (1.21) are acceptable, ultimately, because (1.22) yields the same number of states per unit volume and per unit energy range – the same density of states – as (1.17). From (1.21), the number of allowed values of k_x in a range δk_x is $\delta n_1 = L\delta k_x/2\pi$, and similarly for δk_y and δk_z. If we think of $\boldsymbol{k}$ as a vector in '$\boldsymbol{k}$-space', with axes along k_x, k_y and k_z, and if we think of $\delta^3\boldsymbol{k} = \delta k_x \delta k_y \delta k_z$ as a volume element in $\boldsymbol{k}$-space, then the number of allowed vectors $\boldsymbol{k}$ in this volume element, i.e. the number of different sets of quantum numbers n_1, n_2, n_3, is given by

$$\delta n_1 \delta n_2 \delta n_3 = (L/2\pi)^3 \delta k_x \delta k_y \delta k_z = (L/2\pi)^3 \delta^3\boldsymbol{k}.$$

But to each set of quantum numbers n_1, n_2, n_3 there correspond two states, one of spin up and one of spin down, so that altogether in volume element $\delta^3\boldsymbol{k}$, the number of states available is

$$\delta n_s = 2\delta n_1 \delta n_2 \delta n_3 = V_r \delta^3\boldsymbol{k}/4\pi^3 \tag{1.23}$$

where $V_r = L^3$ is the volume of the cube in 'real space'. This is a very basic result, which can be shown to hold not just for a cube but for a body of any shape; as we shall see later, it also holds not just for the free-electron model but for electrons moving through an ionic lattice too.

For free electrons, the energy $\varepsilon(\boldsymbol{k})$ is related to the wave-vector $\boldsymbol{k}$ by

$$\varepsilon = \hbar^2 k^2/2m \tag{1.24}$$

(where $k = |\boldsymbol{k}|$), as we find at once by inserting (1.20) into (1.16). (The modulus of $\boldsymbol{k}$ should not be confused with Boltzmann's constant, which almost always appears in the combination kT. We normally use the same symbol k for both, but where confusion might arise we shall write Boltzmann's constant as k_{b}.) Thus ε depends only on the magnitude k of the wave-vector, so that all states of a given energy ε will lie on a spherical surface in $\boldsymbol{k}$-space, of radius $k = (2m\varepsilon)^{1/2}/\hbar$. States with energies between ε and $\varepsilon + \delta\varepsilon$ will lie in the region between the spheres of radii k and $k + \delta k$, of volume $4\pi k^2 \delta k$, where

$$\delta k = (\mathrm{d}k/\mathrm{d}\varepsilon)\delta\varepsilon = (2m)^{1/2}\delta\varepsilon/2\hbar\varepsilon^{1/2} \tag{1.25}$$

Thus the number of states available per unit volume between ε and $\varepsilon + \delta\varepsilon$, $g(\varepsilon)\delta\varepsilon$ say, will be given by

$$g(\varepsilon)\delta\varepsilon = \frac{1}{4\pi^3} 4\pi k^2 \delta k = \frac{1}{2\pi^2}\left(\frac{2m}{\hbar^2}\right)^{3/2} \varepsilon^{1/2}\delta\varepsilon \tag{1.26}$$

from (1.23), (1.24) and (1.25). The *density of states* $g(\varepsilon)$ thus varies, for free electrons, as $\varepsilon^{1/2}$.

Not all these states are occupied, of course. At $T = 0$, all states up to the Fermi energy $\varepsilon_{\mathrm{F},0}$ are occupied, and all higher states are empty. The total number of electrons per unit volume, $N/V = n$ (the number density) is then given by

$$n = \int_0^{\varepsilon_{\mathrm{F},0}} g(\varepsilon)\,\mathrm{d}\varepsilon \tag{1.27}$$

or for free electrons, using $g(\varepsilon) \propto \varepsilon^{1/2}$,

$$n = 2g(\varepsilon_{\mathrm{F},0})\varepsilon_{\mathrm{F},0}/3 = \frac{1}{3\pi^2}\left(\frac{2m\varepsilon_{\mathrm{F},0}}{\hbar^2}\right)^{3/2} \tag{1.28}$$

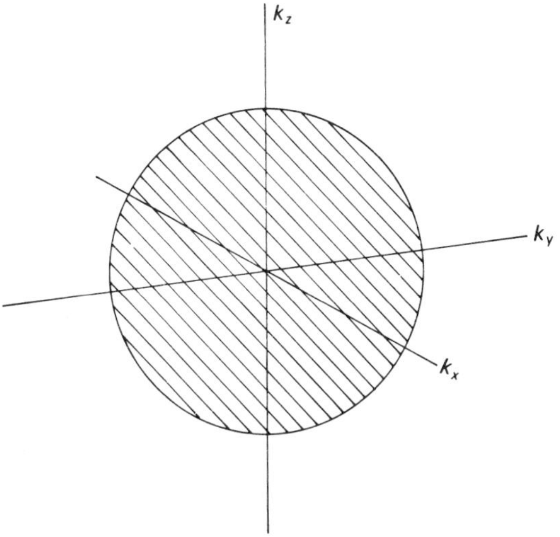

Fig. 1.5 The Fermi sphere of filled states in $\boldsymbol{k}$-space at $T=0$.

(This also follows at once from $n = V_k/4\pi^3$, putting $V_k = 4\pi k_{\mathrm{F},0}^3/3$ and $k_{\mathrm{F},0} = (2m\varepsilon_{\mathrm{F},0}/\hbar^2)^{1/2}$.) The Fermi energy $\varepsilon_{\mathrm{F},0}$ is therefore given by

$$\varepsilon_{\mathrm{F},0} = \hbar^2(3\pi^2 n)^{2/3}2m \tag{1.29}$$

All states in $\boldsymbol{k}$-space inside the spherical Fermi surface (FS)

$$k = k_{\mathrm{F},0} = (3\pi^2 n)^{1/3} \tag{1.30}$$

are full, and all states outside this surface are empty (Fig. 1.5).

How big are $\varepsilon_{\mathrm{F},0}$ and $k_{\mathrm{F},0}$ for a real metal? If the number of conduction electrons is comparable with the number of positive ions, we shall have $n \sim 10^{29}\,\mathrm{m}^{-3}$, so that from (1.30) $k_{\mathrm{F},0} \sim 10^{10}\,\mathrm{m}^{-1}$. This corresponds to a Fermi energy $\varepsilon_{\mathrm{F},0}$ of several eV, and a Fermi velocity $v_{\mathrm{F},0} = \hbar k_{\mathrm{F},0}/m$ of $\sim 10^6\,\mathrm{m\,s}^{-1}$. Thus the electrons at the FS, even at $T=0$, are travelling about ten times as fast as a classical electron would travel at 300 K. And since $kT = 0.025$ eV at 290 K, we have $kT/\varepsilon_{\mathrm{F}} \ll 1$, and the electron gas is indeed strongly degenerate.

2

Properties of free-electron solids

2.1 THE ELECTRONIC HEAT CAPACITY

In this chapter, we shall look at the thermal and electrical properties of the Sommerfeld free-electron model, starting with the electronic heat capacity.

Both ε_F and the total energy U of the electron gas vary (slightly) with temperature, and we look first at ε_F. For $T > 0$, the probability of a state being occupied is given by the FD function $f_0(\varepsilon)$, so that the FS is no longer completely sharp, but blurred out a little. The number density is now given by

$$n = \int_0^\infty g(\varepsilon) f_0(\varepsilon)\, d\varepsilon \tag{2.1}$$

corresponding to the shaded area in Fig. 2.1. Now if n is to be the same as at $T = 0$, the shaded area must be the same as the $T = 0$ area OAB; that is, the areas marked 1 and 2 must be equal. To achieve this ε_F will have to fall slightly from its $T = 0$ value $\varepsilon_{F,0}$, as shown by the dashed line. How much it falls by is given by (1.15). We have

$$n = \int_0^\infty g(\varepsilon) f_0(\varepsilon)\, d\varepsilon = \int_0^{\varepsilon_F} g(\varepsilon)\, d\varepsilon + \frac{\pi^2}{6} k^2 T^2 \left[\frac{dg(\varepsilon)}{d\varepsilon}\right]_{\varepsilon_F}$$

Equating this with the value of n given by (1.27), we have

$$\frac{\pi^2}{6} k^2 T^2 \left[\frac{dg(\varepsilon)}{d\varepsilon}\right]_{\varepsilon_F} = \int_{\varepsilon_F}^{\varepsilon_{F,0}} g(\varepsilon)\, d\varepsilon \approx (\varepsilon_{F,0} - \varepsilon_F) g(\varepsilon_{F,0}) \tag{2.2}$$

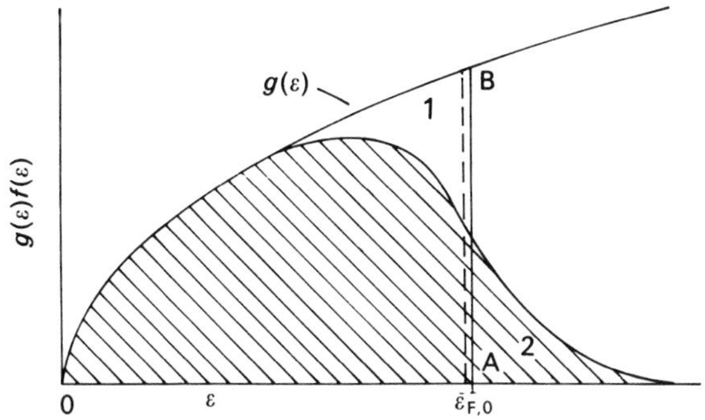

Fig. 2.1 At finite temperatures, the blurring of the FD function leads to a downward shift of ε_F, and to the excitation of electrons from states below ε_F, (1), to states above ε_F, (2).

if $\varepsilon_{F,0} - \varepsilon_F$ is small. Now from (1.26), $[dg(\varepsilon)/d\varepsilon]_{\varepsilon_F} = g(\varepsilon_F)/2\varepsilon_F$ for free electrons, so that

$$(\varepsilon_{F,0} - \varepsilon_F)/\varepsilon_F = (\pi kT/\varepsilon_F)^2/12$$

Since $\varepsilon_F/k \sim 10^4$–$10^5$ K for a typical metal, this shows that the fractional change in ε_F with temperature will be very small.

The thermal excitation of electrons from states below the FS to states above it will also lead to an increase in the total energy of the electrons. We can readily estimate this, using Fig. 2.1. The number of electrons thermally excited is given by the area of region 1: roughly, a triangle of width $\sim 2kT$, height $g(\varepsilon_F)/2$ and area $kTg(\varepsilon_F)/2$. Each electron has gained, on average, an energy of about $2kT$, so that the total increase of energy (per unit volume of metal) is $\Delta U \sim (kT)^2 g(\varepsilon_F)$. We thus expect an electronic contribution to the heat capacity of $C_{el} = d(\Delta U)/dT \sim 2k^2 T g(\varepsilon_F)$.

For a more precise estimate, we again use (1.15). The total energy of the electrons is given by

$$U = \int_0^\infty \varepsilon g(\varepsilon) f_0(\varepsilon)\, d\varepsilon = \int_0^{\varepsilon_F} \varepsilon g(\varepsilon)\, d\varepsilon + \frac{\pi^2}{6} k^2 T^2 \left\{ g(\varepsilon_F) + \varepsilon_F \left[\frac{dg(\varepsilon)}{d\varepsilon}\right]_{\varepsilon_F} \right\}$$

Now the first term on the right can be written

$$\int_0^{\varepsilon_{F,0}} \varepsilon g(\varepsilon)\, d\varepsilon - \int_{\varepsilon_F}^{\varepsilon_{F,0}} \varepsilon g(\varepsilon)\, d\varepsilon = U_0 - (\varepsilon_{F,0} - \varepsilon_F)\varepsilon_F g(\varepsilon_F)$$

(where $U_0 = U(T=0)$),

$$= U_0 - \frac{\pi^2}{6} k^2 T^2 \varepsilon_F \left[\frac{dg(\varepsilon)}{d\varepsilon}\right]_{\varepsilon_F}$$

using (2.2). We thus find $U = U_0 + \pi^2(kT)^2 g(\varepsilon_F)/6$, and

$$C_{el} = dU/dT = \pi^2 k^2 T g(\varepsilon_F)/3 \tag{2.3}$$

For free electrons, (1.28) then gives

$$C_{el} = \pi^2 nk(kT/\varepsilon_F)/2, \tag{2.4}$$

far smaller than the classical value $3nk/2$. The outstanding difficulty of the Drude theory is thus resolved. Experiment confirms that metals do have a linear term in the heat capacity (usually written γT), and for most metals its magnitude is given, to within a factor two or so, by (2.4). In fact (2.2) and (2.3) are valid not only for free electrons but also for electrons moving through an ionic lattice, for which $g(\varepsilon)$ does not have the form (1.26). Measurements of γ thus offer a way of finding the density of states $g(\varepsilon_F)$.

Except at low temperatures, the small γT term is swamped by the much larger heat capacity due to thermal vibrations of the ions of the crystal lattice. But at low temperatures the lattice term vanishes as T^3, and in metals the γT term then dominates. Experimentally, γ is usually found as the intercept in a plot of C/T against T^2, as shown in Fig. 2.2.

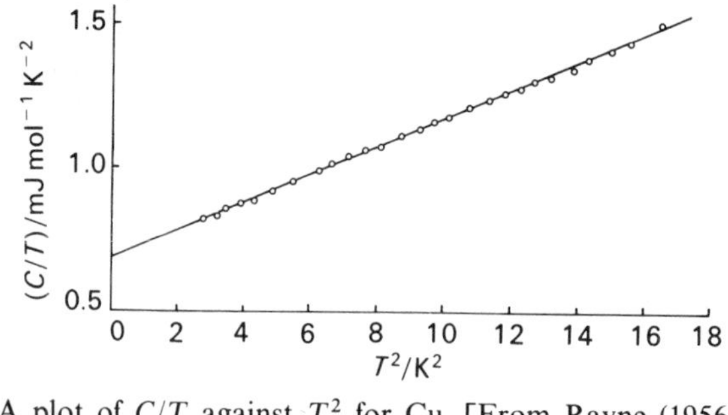

Fig. 2.2 A plot of C/T against T^2 for Cu. [From Rayne (1956), *Aust. J. Phys.* **9**, 189.]

2.2 THE MAGNETIC SUSCEPTIBILITY

The magnetic susceptibility of an electron gas never presented particular problems to the Drude theory, but only because the electron spin and the associated magnetic moment were not discovered until 1925. In 1926 Pauli pointed out that in a classical electron gas, the magnetic moments of the electrons would give rise to a far larger magnetic susceptibility than was observed experimentally. In a magnetic field $\boldsymbol{B}$, an electron spin pointing against the field could lower its energy by turning round to point along the field, and if it were not for the exclusion principle there would be nothing but thermal agitation to prevent all the spins lining up parallel to the field. The relative probability of the two spin orientations would be given by $\exp \pm \mu_B B/kT$, where μ_B is the magnetic moment of the electron, and it is easily shown that if $\mu_B B \ll kT$ the susceptibility would be given by

$$\chi = n\mu_B^2/kT \tag{2.5}$$

Just as with the heat capacity, it is the exclusion principle which causes the observed effect to be much smaller than this classical prediction, as Pauli showed. In a degenerate electron gas, both the spin-up and spin-down states are fully occupied up to energy ε_F, so that the spin-down electrons cannot just reverse their spins – all the spin-up states below ε_F are already fully occupied. The only spin-down electrons that can change to spin-up are those with energies very close to ε_F, as shown in Fig. 2.3. This shows separately the occupied states for spin-up and for spin-down electrons (at $T=0$), after a field $\boldsymbol{B}$ has been applied, but before any spin reversals have taken place. The spin-down states on the left have all been increased in energy by $\mu_B B$, and the spin-up states on the right have been lowered by the same amount. Clearly, the spin-down electrons above ε_F can now lower their energies by occupying the empty spin-up states below ε_F, and this is what happens. The number of electrons involved will be $\mu_B B g(\varepsilon_F)/2$ (since the density of states for each spin is $g(\varepsilon_F)/2$), and the resultant magnetization will be $\mu_B^2 B g(\varepsilon_F)$ per unit volume, since each spin reversal contributes $2\mu_B$ to the magnetization. We thus end up with a weak, temperature-independent Pauli spin susceptibility

$$\chi_P = \mu_B^2 g(\varepsilon_F) \tag{2.6}$$

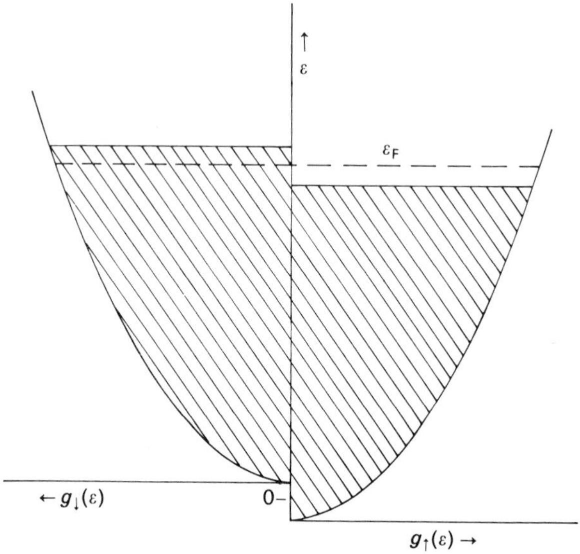

Fig. 2.3 Showing the origin of the Pauli spin susceptibility. The spin-down states on the left have their energy increased and the spin-up states on the right have their energy decreased by $\mu_B B$.

far smaller than (2.5), and in far better agreement with experiment.

It is not easy to measure χ_P experimentally, because there are other contributions to χ of similar magnitude from the ion cores and from the orbital motion of the conduction electrons (section 7.3), but in a few cases χ_P has been measured indirectly by spin-resonance techniques, and the results agree with (2.6) to within a factor of two.

2.3 TRANSPORT PROPERTIES

In section 1.1, we briefly outlined the Drude treatment of electrical and thermal conductivity, and the Sommerfeld treatment is very similar. We think of the electrons as localized wave-packets, not very different from point particles except that they obey the exclusion principle, and consequently obey FD statistics rather than Boltzmann statistics. Because of this, the transport properties of a metal are governed entirely by the electrons near the Fermi surface. The states well below the FS are completely full, and those well above the FS are completely empty, so that it is only electrons in states close to the FS that are able to respond to applied fields.

Just as in section 1.1, we can argue that if the electrons at the FS have a relaxation time τ, they will acquire an average drift velocity $\boldsymbol{v}_d = -e\boldsymbol{E}\tau/m$ in an applied field $\boldsymbol{E}$. Now since for free electrons $m\boldsymbol{v} = \hbar\boldsymbol{k}$, this means that each electron at the FS will have its $\boldsymbol{k}$-vector changed by an amount $\boldsymbol{k}_d = m\boldsymbol{v}_d/\hbar$. Since all states below the FS are filled, this means in effect that the whole electron gas – the whole sphere of filled states in $\boldsymbol{k}$-space – has been shifted by this amount, so that in effect all the electrons in the metal have acquired the same drift velocity $\boldsymbol{v}_d$, though the value of $\boldsymbol{v}_d$ is determined by the electrons at the FS. It follows that the electrical conductivity σ is given by precisely the same expression as on the Drude model,

$$\sigma = ne^2\tau/m \tag{2.7}$$

or

$$= ne^2 l/mv_F \tag{2.8}$$

since $l = \tau v_F$ for the electrons at the FS.

To find the thermal conductivity, we again use the kinetic theory expression (1.8), but with $\bar{v}$ now replaced by v_F, because κ, like σ, is now determined by the electrons at the FS. If we also use (2.4) for the electronic heat capacity, we find

$$\begin{aligned} \kappa &= \pi^2 nk^2 T v_F l/6\varepsilon_F \\ &= (\pi^2/3)nk^2\tau T/m, \end{aligned} \tag{2.9}$$

using $l = \tau v_F$ and $\varepsilon_F = \frac{1}{2}mv_F^2$. The Lorenz number is thus given by

$$\mathscr{L} = \kappa/\sigma T = (\pi^2/3)(k/e)^2 = 2.45 \times 10^{-8}\,\mathrm{V^2\,K^{-2}} \tag{2.10}$$

in excellent agreement with experiment. The average experimental value of $\mathscr{L}$ for 24 reasonably good metallic conductors is $2.42 \times 10^{-8}\,\mathrm{V^2\,K^{-2}}$ at 273 K, and 2.47 at 373 K. (In poor metals, with low conductivities, there is an appreciable additional contribution to κ from the heat conducted by the ionic lattice itself, which increases $\mathscr{L}$ above the theoretical value.)

The Sommerfeld theory thus accounts for the Wiedemann–Franz law even better than the original Drude theory. We still have to explain why the experimental value of $\mathscr{L}$ falls off below room temperature (Fig. 1.3), and we shall see later that this is because the effective values of τ to be used in (2.7) and (2.9) then become rather different.

What values of τ, and of l, does the Sommerfeld theory imply? The values of τ derived from σ will be the same as on the Drude theory, because expressions (1.5) and (2.7) are identical in form. But the values of l will be much larger, because v_F in a metal, even at $T = 0$, is an order of magnitude larger than the classical velocity v_r would be, even at 300 K. And since v_F does not vary appreciably with temperature, the whole of the temperature-dependence of σ in (2.8) must now be attributed to variation of l. Thus for Cu at room temperature, we find $l \sim 40$ nm, and in reasonably pure samples at low temperatures, σ can rise by a factor of 100 or more, so that l must rise to 4 μm or more. In some extremely pure metal samples, l rises to 1 mm or more at very low temperatures.

Such long mean free paths would have been completely incomprehensible classically, in terms of a negatively charged electron fighting its way through an array of positively charged ions. As we shall see later, quantum mechanics resolves the problem: in a perfect crystal, an electron would never collide at all, and the collisions that occur in real crystals happen because the crystal is less than perfect. In reasonably pure metals, the main 'imperfections' are just the thermal vibrations of the ions, which grow in amplitude as the temperature is raised, so that l falls and $\rho = 1/\sigma$ rises (Fig. 1.1). At low enough temperatures, the scattering of electrons by the ionic vibrations becomes negligible, and all that is left is scattering due to static imperfections in the crystal such as dislocations, vacancies, grain boundaries and impurity atoms (section 8.1). This residual scattering causes a temperature-independent residual resistance which varies from sample to sample; in a very pure, carefully grown single crystal it may be extremely small.

If the thermal vibrations cause an electron to collide $1/\tau_t$ times per second, and the residual imperfections cause it to collide $1/\tau_r$ times per second, the total scattering rate will be $1/\tau = 1/\tau_t + 1/\tau_r$ if the two separate scattering rates are simply additive. Since $\rho \propto 1/\tau$, we can then write

$$\rho(T) = \rho_t(T) + \rho_r \tag{2.11}$$

where ρ_t should depend only on temperature and not on sample perfection, while ρ_r depends only on sample perfection and not on temperature. The imperfections which determine ρ_r include both impurities and physical imperfections such as vacancies, etc. They must not be too numerous or they will begin to affect the resistivity in other ways too, for example by altering the number of conduction electrons or by affecting the amplitude of the ionic vibrations and so changing ρ_t. But for impurity concentrations not exceeding a few per cent, the additivity predicted by (2.11) is quite well obeyed experimentally, and was discovered by Matthiessen in 1864. It implies that the $\rho(T)$ curves of two samples of the same metal, differing only in sample perfection, should run parallel to each other, as shown in Fig. 1.1. It is a little surprising, in fact, that Matthiessen's rule is obeyed as closely as it is. The argument leading to (2.11) is based on a very simplified model of a metal, and on any more realistic model we would not expect (2.11) to be obeyed exactly.

2.4 HALL EFFECT AND MAGNETORESISTANCE

When a metal (or a semiconductor) carrying a current of density $\boldsymbol{J}$ is placed in a magnetic field $\boldsymbol{B}$, normal to $\boldsymbol{J}$, an electric field $\boldsymbol{E}_H$ is set up, normal to both $\boldsymbol{B}$ and $\boldsymbol{J}$. This effect was discovered by Hall in 1879. At the same time, the field $\boldsymbol{E}_J$ parallel to $\boldsymbol{J}$ has to be increased, if $\boldsymbol{J}$ is to be kept unchanged; in other words, the resistivity $\rho = \boldsymbol{E}_J/\boldsymbol{J}$ has been increased by the field $\boldsymbol{B}$. This effect is called transverse magnetoresistance – transverse, because $\boldsymbol{B}$ is normal to $\boldsymbol{J}$. A similar but smaller rise in the resistivity (longitudinal magnetoresistance) occurs if $\boldsymbol{B}$ is parallel to $\boldsymbol{J}$, but in that case there is no Hall field $\boldsymbol{E}_H$.

Can the Sommerfeld theory explain these results? Consider first what happens to a free-electron metal when we place it in a field $\boldsymbol{B}$, with no applied field $\boldsymbol{E}$, so that $\boldsymbol{J} = 0$. An electron of charge $-e$, moving with velocity $\boldsymbol{v}$, will be subject to a Lorentz force $\boldsymbol{F} = -e\boldsymbol{v} \times \boldsymbol{B}$, and $\boldsymbol{v}$ will therefore change at a rate given by

$$m\dot{\boldsymbol{v}} = -e\boldsymbol{v} \times \boldsymbol{B} \tag{2.12}$$

If $\boldsymbol{B} = (0, 0, B_z)$, we have $\boldsymbol{v} \times \boldsymbol{B} = (v_y B_z, -v_x B_z, 0)$, so that

$$\dot{v}_x = -(e/m)v_y B_z, \quad \dot{v}_y = (e/m)v_x B_z, \quad \dot{v}_z = 0 \tag{2.13}$$

Thus $v_z =$ constant, and is unaffected by the field B_z. To solve the equations in v_x and v_y, write $v_t = v_x + iv_y$ (where the subscript 't' denotes 'transverse'); we can then combine the equations for v_x and v_y into the single equation

$$\dot{v}_t = i\omega_c v_t \tag{2.14}$$

where $\omega_c = eB_z/m$. Clearly, the solution of (2.14) is

$$v_t = v_{t,0}\, e^{i(\omega_c t + \phi)}$$

or

$$v_x = v_{t,0} \cos(\omega_c t + \phi), \quad v_y = v_{t,0} \sin(\omega_c t + \phi) \tag{2.15}$$

where $v_{t,0}$ and ϕ are constants. If we think of $\boldsymbol{v}$ as a point in 'velocity space', (2.15) tells us that $\boldsymbol{v}$ moves in a circle around the z axis at the angular frequency ω_c, known as the cyclotron frequency. The free-electron gas as a whole will be represented by a filled sphere of points in $\boldsymbol{v}$-space, of radius v_F, and the whole sphere will be rotating at ω_c.

Since for free electrons $\boldsymbol{v} = \hbar\boldsymbol{k}/m$, we can equally well describe the situation by saying that the whole Fermi sphere of occupied states in $\boldsymbol{k}$-space is likewise rotating at ω_c. Because the occupied states in $\boldsymbol{k}$-space are the same set of lowest-energy states as in $\boldsymbol{B} = 0$, collisions can do nothing to reduce their energy further, and therefore collisions have no effect on the solution.

Suppose now that we apply a field $\boldsymbol{E}$ as well as $\boldsymbol{B}$, so that (2.12) becomes

$$m\dot{\boldsymbol{v}} = -e(\boldsymbol{E} + \boldsymbol{v} \times \boldsymbol{B}) \tag{2.16}$$

As usual, $\boldsymbol{E}$ will produce a drift velocity $\boldsymbol{v}_d$ of the electron gas, and as usual collisions will play a part in determining $\boldsymbol{v}_d$, but now so will $\boldsymbol{B}$. To solve (2.16) in the presence of collisions, we use a device which has the merit of giving the right answer quite simply, even if it is a little difficult to justify rigorously. We write $\boldsymbol{v} = \boldsymbol{v}_c + \boldsymbol{v}_d$, where $\boldsymbol{v}_c$ satisfies

$$m\dot{\boldsymbol{v}}_c = -e(\boldsymbol{v}_c \times \boldsymbol{B}), \tag{2.17}$$

i.e. the same equation as (2.12), with the same solutions. Subtracting

(2.17) from (2.16), we are left with

$$m\dot{\boldsymbol{v}}_{\mathrm{d}} = -e(\boldsymbol{E} + \boldsymbol{v}_{\mathrm{d}} \times \boldsymbol{B}) \tag{2.18}$$

We now suppose, as in $\boldsymbol{B} = 0$, that the accelerating force on the rhs acts on average for a time τ to produce the steady drift velocity

$$\boldsymbol{v}_{\mathrm{d}} = -e\tau(\boldsymbol{E} + \boldsymbol{v}_{\mathrm{d}} \times \boldsymbol{B})/m \tag{2.19}$$

To solve (2.19), consider two alternative situations: (i) $\boldsymbol{v}_{\mathrm{d}} \parallel \boldsymbol{B}$; (ii) $\boldsymbol{v}_{\mathrm{d}} \perp \boldsymbol{B}$. In case (i), $\boldsymbol{v}_{\mathrm{d}} \times \boldsymbol{B} = 0$, and we at once find $\boldsymbol{v}_{\mathrm{d}} = -e\boldsymbol{E}\tau/m$, so that $\boldsymbol{E} \parallel \boldsymbol{v}_{\mathrm{d}}$, and this is thus the solution for $\boldsymbol{E}$ parallel to $\boldsymbol{B}$. Since $\boldsymbol{J} = -ne\boldsymbol{v}_{\mathrm{d}}$, we at once regain the usual result $\boldsymbol{J} = \sigma\boldsymbol{E}$, with $\sigma = ne^2\tau/m$. So for $\boldsymbol{E} \parallel \boldsymbol{B}$, the presence of $\boldsymbol{B}$ has no effect on the conductivity at all: the Sommerfeld model gives no longitudinal magnetoresistance.

In case (ii), suppose again that $\boldsymbol{B} = B_z$, and that $\boldsymbol{E} = E_x$. We then have

$$v_{\mathrm{d},x} = -e\tau(E_x + v_{\mathrm{d},y}B_z)/m, \quad v_{\mathrm{d},y} = e\tau v_{\mathrm{d},x}B_z/m,$$

and if as before we write $v_{\mathrm{d,t}} = v_{\mathrm{d},x} + iv_{\mathrm{d},y}$, we can combine these equations into

$$v_{\mathrm{d,t}} = -e\tau(E_x - iB_z v_{\mathrm{d,t}})/m$$

with solution

$$v_{\mathrm{d,t}} = -e\tau E_x/m(1 - i\omega_{\mathrm{c}}\tau)$$

Thus

$$J_{\mathrm{t}}/E_x = ne^2\tau/m(1 - i\omega_{\mathrm{c}}\tau) = \sigma_0/(1 - i\omega_{\mathrm{c}}\tau) \tag{2.20}$$

where $J_{\mathrm{t}} = J_x + iJ_y = -nev_{\mathrm{d,t}}$ and $\sigma_0 = ne^2\tau/m$ as usual. Taking the real and imaginary parts of (2.20), and writing $J_x = \sigma_{xx}E_x$, $J_y = \sigma_{yx}E_x$, we find

$$\sigma_{xx} = \sigma_0/(1 + \omega_{\mathrm{c}}^2\tau^2), \quad \sigma_{yx} = \sigma_0\omega_{\mathrm{c}}\tau/(1 + \omega_{\mathrm{c}}^2\tau^2) \tag{2.21}$$

A transverse magnetic field, $\boldsymbol{B} \perp \boldsymbol{E}$, thus has a very marked effect on the current $\boldsymbol{J}$ flowing in the direction of $\boldsymbol{E}$, and also produces a current normal to both $\boldsymbol{B}$ and $\boldsymbol{E}$.

But in practice, we do not apply a field $\boldsymbol{E}$ to a metal sample, and then look to see the direction in which $\boldsymbol{J}$ flows; we send a current of density $\boldsymbol{J}$ in a particular direction – along the wire – and look to see what field $\boldsymbol{E}$ is needed in the direction of $\boldsymbol{J}$, and what field $\boldsymbol{E}_{\mathrm{H}}$ appears transverse to $\boldsymbol{J}$. Either way, what we are really concerned with is the relation between the magnitude and direction of the two vectors $\boldsymbol{E}$ and $\boldsymbol{J}$ in the xy plane, and either way this is given by (2.20). If we invert (2.20) and choose the direction of $\boldsymbol{J}$ as the x axis, we have

$$E_{\mathrm{t}}/J_x = (E_x + iE_y)/J_x = \rho_0(1 - i\omega_{\mathrm{c}}\tau) \tag{2.22}$$

where $\rho_0 = 1/\sigma_0$ is the resistivity in $\boldsymbol{B} = 0$. Writing $E_x = \rho_{xx} J_x$, $E_y = \rho_{yx} J_x$, we find

$$\rho_{yx} = -\rho_0\omega_{\mathrm{c}}\tau = -(m/ne^2\tau)(eB_z/m)\tau = -B_z/ne; \quad \rho_{xx} = \rho_0. \tag{2.23}$$

On the Sommerfeld model, then, the measured resistivity ρ_{xx} is *independent* of $\boldsymbol{B}$: the transverse magnetoresistance vanishes, like the longitudinal magnetoresistance, but we do find a Hall field $E_y = -B_zJ_x/ne$. The ratio E_y/B_zJ_x is called the Hall coefficient R_{H}, and for this model we find

$$R_{\mathrm{H}} = -1/ne \tag{2.24}$$

It may seem surprising at first that (2.21) and (2.23) do indeed

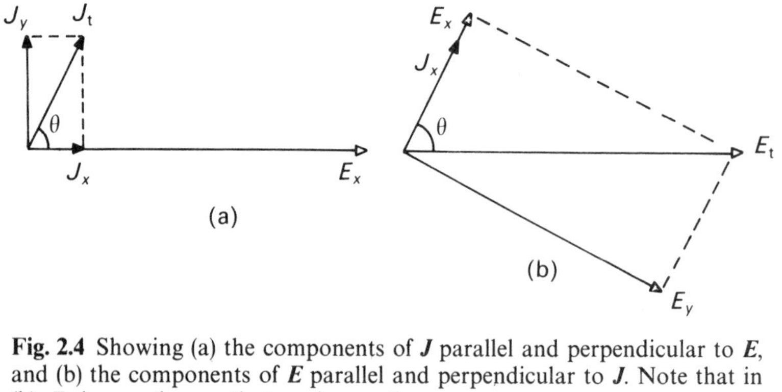

Fig. 2.4 Showing (a) the components of $\boldsymbol{J}$ parallel and perpendicular to $\boldsymbol{E}$, and (b) the components of $\boldsymbol{E}$ parallel and perpendicular to $\boldsymbol{J}$. Note that in (b), E_y is negative, so that ρ_{yx} is negative.

describe the same physical behaviour: at large fields B_z, σ_{xx} tends to zero as $1/B_z^2$, whereas ρ_{xx} is completely field-independent. Figure 2.4 shows how this happens, and should make clear the relationship between the two descriptions.

Experimentally, (2.24) succeeds very well in predicting R_H for the alkali metals, and reasonably well for the noble metals (Cu, Ag, Au), if n is set equal to n_a, the number of atoms per unit volume. For Cu, for example, (2.24) predicts $R_\mathrm{H} = 7.4 \times 10^{-11}\,\mathrm{m^3\,C^{-1}}$, and measurements give $5.0 \times 10^{-11}\,\mathrm{m^3\,C^{-1}}$ at 300 K, rising to about 6.2×10^{-11} at low temperatures and high fields. But for many other metals, although (2.24) predicts about the right order of magnitude for R_H, it predicts the wrong sign: the Hall coefficient is positive instead of negative. As we shall see later, the explanation is that conduction electrons moving through an ionic lattice can sometimes behave like 'positive holes'. But in terms of the free-electron model, such behaviour is quite inexplicable.

2.5 RELAXATION EFFECTS: HIGH-FREQUENCY CONDUCTIVITY

So far we have assumed the field $\boldsymbol{E}$ to be time-independent. What happens if $\boldsymbol{B} = 0$ but $\boldsymbol{E}$ oscillates at some high frequency, so that $\boldsymbol{E} = \boldsymbol{E}_0\,\mathrm{e}^{i\omega t}$? If we applied very naïvely the argument leading to (1.3) and (2.7), we might write $\boldsymbol{v}_\mathrm{d} = -(e/m)\int_{t-\tau}^{t} \boldsymbol{E}\,\mathrm{d}t$, but then $\boldsymbol{v}_\mathrm{d}$ (and hence $\boldsymbol{J}$) would vanish whenever the frequency satisfied $\omega\tau = 2n\pi$, which is obviously implausible. In fact, of course, not all the electrons have travelled for exactly the same time τ since their last collision, and in fact the times for which they have travelled are exponentially distributed. A fraction $\mathrm{e}^{-u/\tau}\,\delta u/\tau$ will have travelled for a time between u and $u + \delta u$, and will have acquired a drift velocity $-(e/m)\boldsymbol{E}\int_0^u \mathrm{e}^{-i\omega s}\,\mathrm{d}s$, because the field at time $t - s$ is $\boldsymbol{E}_0\,\mathrm{e}^{i\omega(t-s)} = \boldsymbol{E}\,\mathrm{e}^{-i\omega s}$. The mean drift velocity is found by averaging this over all u:

$$\boldsymbol{v}_\mathrm{d} = -(e\boldsymbol{E}/m\tau)\int_0^\infty \mathrm{d}u\,\mathrm{e}^{-u/\tau}\int_0^u \mathrm{e}^{-i\omega s}\,\mathrm{d}s \qquad (2.25)$$

On carrying out the integrations (problem 2.4), we find

$$\boldsymbol{v}_\mathrm{d} = -e\boldsymbol{E}\tau/m(1 + i\omega\tau) \qquad (2.26)$$

so that

$$\sigma_\omega = \boldsymbol{J}/\boldsymbol{E} = -ne\boldsymbol{v}_\mathrm{d}/\boldsymbol{E} = \sigma_0/(1 + i\omega\tau) \tag{2.27}$$

where $\sigma_0 = ne^2\tau/m$ as usual.

The result (2.27) looks very much like (2.20), but there the complex conductivity signified an angle between $\boldsymbol{J}$ and $\boldsymbol{E}$ in the xy plane; here it signifies a phase angle in time between $\boldsymbol{J}$ and $\boldsymbol{E}$. The two vectors point in the same direction in space, but as $\omega\tau$ increases $\boldsymbol{J}$ lags more and more behind $\boldsymbol{E}$, until for $\omega\tau \gg 1$ the phase lag approaches 90°. In this limit, $\sigma_\omega \rightarrow -ine^2/m\omega$, independent of τ. The response of the electron gas to $\boldsymbol{E}$ is then limited not by collisions but by the inertia of the electrons.

At room temperature, $\tau \approx 10^{-14}$ s for typical metals, so that $\omega\tau$ becomes appreciable, and 'relaxation' effects become important, only at infra-red frequencies and above. At these frequencies, a metal is characterized experimentally by its reflectivity rather than its conductivity. The reflectivity is determined by the refractive index N, which is in turn determined by the relative dielectric constant ε_r and the conductivity σ of the medium. (As always, it is difficult to find enough different symbols to denote all the different things we want to label. Note that in this section, ε and N do not mean energy or number of particles.)

Suppose first that $\sigma = 0$. Then Maxwell's equations show that an electromagnetic wave travels through the medium at speed c/N, where $N = \varepsilon_\mathrm{r}^{1/2}$, so that the wave has the form

$$\boldsymbol{E} = \boldsymbol{E}_0 \exp i\omega(t - Nx/c) \tag{2.28}$$

if it is travelling in the x direction. If a wave of amplitude $\boldsymbol{E}_\mathrm{i}$ is incident normally on the surface of the medium at $x = 0$, the reflected and transmitted amplitudes are found by matching the total electric fields and the total magnetic fields across the boundary, and turn out to be given by

$$\boldsymbol{E}_\mathrm{r}/\boldsymbol{E}_\mathrm{i} = (1 - N)/(1 + N), \quad \boldsymbol{E}_\mathrm{t}/\boldsymbol{E}_\mathrm{i} = 2/(1 + N). \tag{2.29}$$

If now $\sigma \neq 0$, the results (2.28) and (2.29) still hold, but we have to replace ε_r by an effective dielectric constant ε_eff, so that

$$N = \varepsilon_\mathrm{eff}^{1/2}. \tag{2.30}$$

To see the form of ε_{eff}, we write the relevant Maxwell equation as

$$\begin{aligned}\operatorname{curl} \boldsymbol{H} &= \boldsymbol{J} + \dot{\boldsymbol{D}} \\ &= \sigma_\omega \boldsymbol{E} + i\omega\varepsilon_r\varepsilon_0 \boldsymbol{E} \\ &= i\omega\varepsilon_0(\varepsilon_r + \sigma_\omega/i\omega\varepsilon_0)\boldsymbol{E}\end{aligned}$$

Thus we can write

$$\operatorname{curl} \boldsymbol{H} = i\omega\varepsilon_0\varepsilon_{\text{eff}}\boldsymbol{E}$$

where

$$\varepsilon_{\text{eff}} = \varepsilon_r + \sigma_\omega/i\omega\varepsilon_0 \tag{2.31}$$

and includes the effects of both ε_r and σ_ω. We have assumed here that $\boldsymbol{D}(=\varepsilon_r\varepsilon_0\boldsymbol{E})$ varies as $e^{i\omega t}$, and that $\boldsymbol{J} = \sigma_\omega \boldsymbol{E}$ with σ_ω given by (2.27) for the free-electron model.

Equations equivalent to (2.29)–(2.31) were first discussed by Drude in 1900, and this treatment is usually referred to as the Drude theory of optical properties. Equations (2.30) and (2.31) show that N is complex if $\sigma_\omega \neq 0$, and if we write $N = n_0 - ik_0$ (using the traditional notation), (2.28) shows that the wave decays exponentially in a distance $c/\omega k_0$.

How do ε_{eff} and N vary with ω? At low enough frequencies, when $\omega\tau \ll 1$, we have $\sigma_\omega \approx \sigma_0$, and at these frequencies the first term in (2.31) can be neglected in metals, so that

$$\varepsilon_{\text{eff}} \approx -i\sigma_0/\omega\varepsilon_0 \tag{2.32}$$

This is just the classical skin effect region, where the wave falls off below the surface in a distance

$$\delta = c/\omega k_0 = c(2\varepsilon_0/\sigma_0\omega)^{1/2} = (2/\mu_0\sigma_0\omega)^{1/2} \tag{2.33}$$

Since $|\varepsilon_{\text{eff}}| \gg 1$, and hence $|N| \gg 1$, equation (2.29) shows that $\boldsymbol{E}_r/\boldsymbol{E}_i \approx -1$, and reflection is practically perfect.

At higher frequencies, when $\omega\tau \gg 1$, we have $\sigma_\omega \approx \sigma_0/i\omega\tau$, from (2.27), so that

$$\varepsilon_{\text{eff}} \approx \varepsilon_r - \sigma_0/\varepsilon_0\tau\omega^2 = \varepsilon_r(1 - \omega_p^2/\omega^2) \tag{2.34}$$

where

$$\omega_p^2 = \sigma_0/\varepsilon_r\varepsilon_0\tau = ne^2/m\varepsilon_r\varepsilon_0 \tag{2.35}$$

Equation (2.34) shows that for $1/\tau \ll \omega < \omega_p$, ε_{eff} is real and negative, so that N is imaginary, and the wave dies away in a distance

$$\delta = c/\omega_p \varepsilon_r^{1/2} \tag{2.36}$$

as long as $\omega_p^2/\omega^2 \gg 1$. For metallic values of n, the 'plasma frequency' ω_p is of order $10^{16}\,s^{-1}$, giving a depth δ of order 200 nm. Since N is (almost) purely imaginary, (2.29) shows that the reflected wave has the same amplitude as the incident wave: there is negligible energy loss. Figure 2.5 shows the measured variation of ε_{eff} (or its real part, $Re(\varepsilon_{eff})$) with light wavelength for Cu, and confirms (2.34).

Finally, for $\omega > \omega_p$, ε_{eff} changes sign, from (2.34); N then becomes real and positive, so that from (2.28) the wave propagates without decay, and the metal becomes transparent.

The Drude theory gives a reasonable account of the observed

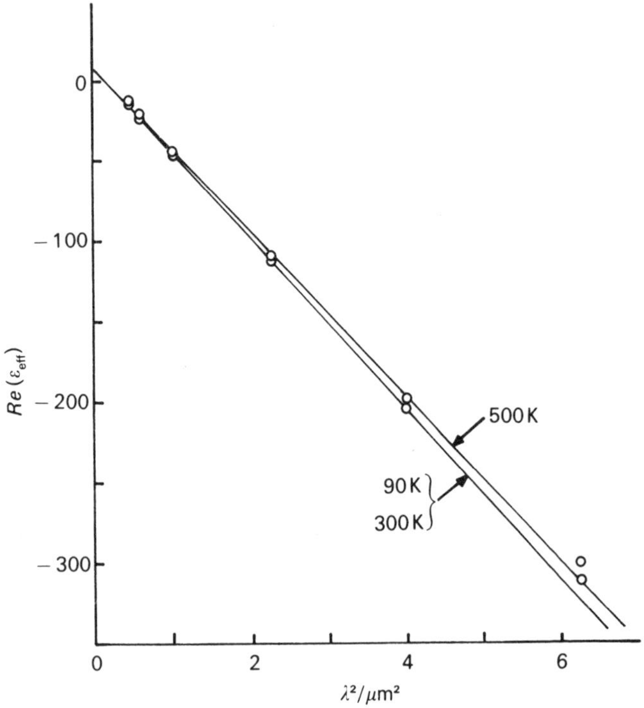

Fig. 2.5 The variation of $Re(\varepsilon_{eff})$ with λ^2 for Cu. [From Roberts (1960), *Phys. Rev.* **118**, 1509.]

behaviour of real metals at room temperature, up to frequencies approaching ω_p. In the alkali metals, indeed, the theory works well right up to $\omega > \omega_p$; these metals show an abrupt fall in reflectivity at a frequency close to the value of ω_p calculated from (2.35), putting $\varepsilon_r = 1$. It is difficult to estimate ε_r precisely – one can think of $\varepsilon_r^{1/2}$ as the refractive index the material *would* have if all the conduction electrons were frozen into place, and unable to move – but it is unlikely to exceed 2 or 3 for most materials.

For most metals other than the alkalis, the Drude theory breaks down for $\omega \sim \omega_p$ or less, and the absorption becomes much greater than Drude predicts, because of 'interband transitions': photons of energy $\hbar\omega$, if ω is high enough, are able to excite electrons to higher-energy states inside the metal, as we shall see later, and this behaviour is of course not included in the classical Drude model, or even in the Sommerfeld model. But at lower frequencies, comparison between the Drude theory and experiment enables estimates to be made of two quantities: the relaxation time τ, because the change-over from (2.32) to (2.34) occurs around $\omega\tau = 1$, and the ratio $n/m\varepsilon_r$ which determines ω_p. The value of ω_p can be found by fitting (2.34) to measurements at lower frequencies, even if (2.34) breaks down at $\omega \sim \omega_p$.

For pure metals at low temperatures, when the mean free path becomes large, the Drude theory breaks down at still lower (microwave) frequencies, not because of quantum effects but because in deriving (2.27) and hence (2.33) we assumed that the field $\boldsymbol{E}$ was spatially uniform over distances of the order of l. Unless the skin depth δ predicted by (2.33) satisfies $\delta \gg l$, this will no longer be valid, and nor will (2.33); classical skin effect theory must be replaced by 'anomalous' skin effect theory, and we return to this in section 12.1.

2.6 METALS AND SEMICONDUCTORS

We have seen that although the classical Drude–Lorentz theory had some serious problems, particularly with the electronic heat capacity, it also had some notable successes, such as the explanation of the Wiedemann–Franz law. The Sommerfeld theory was very similar in approach to the Drude–Lorentz theory, and differed from it only in assuming that the electron gas obeyed FD statistics rather than classical Boltzmann statistics. This at once solved the heat capacity problem, and led to a rather different value for the Lorenz number, which agreed rather better with experiment.

In this chapter, we have discussed the Sommerfeld theory mainly as applied to metals, but it is equally applicable to semiconductors. There are two essential differences between metals and semiconductors. First, in metals the number density of conduction electrons n is large enough for the electron gas to be degenerate, with $\varepsilon_F \gg kT$, whereas in semiconductors n is much smaller, so that the FD function f_0 satisfies $f_0 \ll 1$ for all states in the conduction band, and ε_F (measured from the bottom of the conduction band) is therefore large and negative: $-\varepsilon_F \gg kT$. The FD function (1.11) then becomes indistinguishable from the Boltzmann exponential function (1.13, 1.14), and the electron gas behaves like a classical perfect gas.

The second essential difference is that in metals n is temperature-independent, whereas in semiconductors n is not only much smaller but also strongly temperature-dependent. We shall see later why this is; essentially it is because the electrons are being thermally excited up into the conduction band (leaving behind, in pure semiconductors, 'positive holes' in a lower-energy valence band), and the number excited falls rapidly to zero at low temperatures.

In applying the Sommerfeld theory to semiconductors, then, we have to allow for the temperature-dependence of n, and we have to allow for the fact that the gas is non-degenerate, and therefore behaves essentially like the classical gas assumed by Drude and Lorentz. Indeed, Sommerfeld theory in this limit is virtually the same as Drude–Lorentz theory. In particular, the electronic heat capacity is given by $C_{el} = 3nk/2$, but this is now no problem, because n is so small for semiconductors that this term is unobservably small. Because the electron gas is now classical, the transport properties are no longer determined entirely by the electrons at the Fermi surface – there *is* no Fermi surface – but have to be calculated by averaging over all electrons, of all energies.

If all the electrons are assumed to have the same relaxation time τ, so that they all acquire the same drift velocity $\boldsymbol{v}_d$ in an applied field, then virtually all the results obtained above for a metal apply unchanged to a semiconductor. In particular, σ is still given by (2.7), the behaviour in a magnetic field is still given by (2.22)–(2.24), and the high-frequency behaviour is still given by (2.27)–(2.36). The thermal properties are changed somewhat, even if τ is the same for all electrons, because these properties involve an average over the *energies* of all the electrons, and in a semiconductor the relevant electrons no longer all have effectively the same energy ε_F, as they do in a metal. Thus, the numerical coefficient in κ and in $\mathscr{L}$

(equations 2.9, 2.10) turns out to be 5/2 instead of $\pi^2/3$. But this is of little interest, because the electronic contribution to the total thermal conductivity in a semiconductor is small compared with the lattice contribution (as with the heat capacity), and cannot easily be measured experimentally.

If, more realistically, τ is assumed to vary with the electron energy ε, the calculation of σ, κ, R_H, etc. will involve averaging over different groups of electrons with different drift velocities. This averaging has to be carried out a little carefully. If $n_v(0)\ \delta^3\boldsymbol{v}$ is the number of electrons (in $\boldsymbol{E}=0$) with velocities in the range $\delta v_x \delta v_y \delta v_z = \delta^3\boldsymbol{v}$ about $\boldsymbol{v}$, so that $n = \int n_v(0)\,\mathrm{d}^3\boldsymbol{v}$, it is tempting to write

$$\boldsymbol{J} = -e\int \boldsymbol{v}_\mathrm{d} n_v(0)\,\mathrm{d}^3\boldsymbol{v} \tag{2.37}$$

But if $\boldsymbol{v}_\mathrm{d}$ depends on $\boldsymbol{v}$, this is incorrect; it overlooks the fact that in a field $\boldsymbol{E}$, n_v itself is altered, from $n_v(0)$ to $n_v(\boldsymbol{E})$ say. The electrons which have velocity $\boldsymbol{v}$ in the field $\boldsymbol{E}$ are the electrons which had velocity $\boldsymbol{v}-\boldsymbol{v}_\mathrm{d}$ in $\boldsymbol{E}=0$, so that

$$n_v(\boldsymbol{E}) = n_v(0) - [\boldsymbol{v}_\mathrm{d}\cdot \mathrm{d}n_v(0)/\mathrm{d}\boldsymbol{v}] \tag{2.38}$$

where $\mathrm{d}n_v(0)/\mathrm{d}\boldsymbol{v} = [\boldsymbol{i}\partial n_v(0)/\partial v_x + \boldsymbol{j}\ldots]$. We thus have

$$\boldsymbol{J} = -e\int \boldsymbol{v} n_v(\boldsymbol{E})\,\mathrm{d}^3\boldsymbol{v} = e\int \boldsymbol{v}[\boldsymbol{v}_\mathrm{d}\cdot \mathrm{d}n_v(0)/\mathrm{d}\boldsymbol{v}]\,\mathrm{d}^3\boldsymbol{v} \tag{2.39}$$

where we have dropped the term in $n_v(0)$, since this contributes nothing to the current: when $\boldsymbol{E}=0$, $\boldsymbol{J}=0$. Equation (2.37) leads to the same result as (2.39) when $\boldsymbol{v}_\mathrm{d}$ is independent of $\boldsymbol{v}$, but not otherwise (problem 2.6).

As one might expect, nothing very new emerges if we assume τ to depend on ε and use (2.39) to evaluate σ (and the corresponding expression for the heat current density to evaluate κ); the resultant expressions look much as before, but with τ replaced by some appropriate average. For example, if we assume $\tau \propto \varepsilon^\nu$ say, it turns out that the Lorenz number becomes

$$\mathscr{L} = (5/2 + \nu)k^2/e^2 \tag{2.40}$$

Thus if we assume a constant mean free path l, as Lorentz did, we

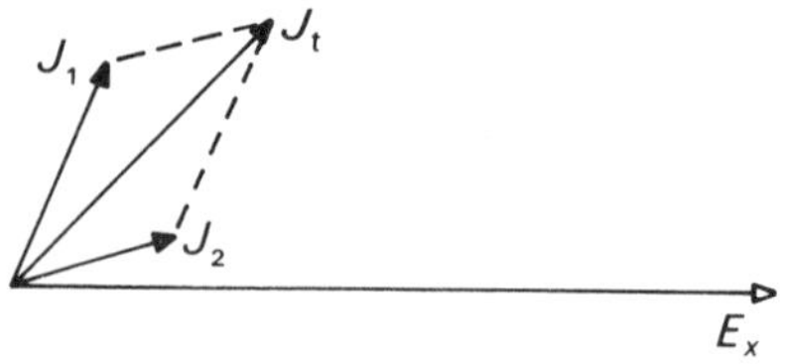

Fig. 2.6 How magnetoresistance can arise, if two groups of electrons have different Hall angles between $\boldsymbol{J}$ and $\boldsymbol{E}$.

have $\tau = l/v$ and so $\tau \propto \varepsilon^{-1/2}$, and we then find, as Lorentz did, that $\mathscr{L} = 2k^2/e^2$.

More interesting is the behaviour in a magnetic field. For $\boldsymbol{E} \parallel \boldsymbol{B}$, nothing new arises; as before, the electrons pursue helical paths around the $\boldsymbol{B}$ direction, but this does not affect their response to a field $\boldsymbol{E}$ in that direction, and the conductivity and resistivity are unaffected by $\boldsymbol{B}$. But for $\boldsymbol{E} \perp \boldsymbol{B}$, the spread in τ does have an effect. If $\delta \boldsymbol{J}$ is the current carried by electrons with energies between ε and $\varepsilon + \delta\varepsilon$, the argument which led (for $\boldsymbol{B} = 0,0,B_z$) to (2.20) now leads to

$$\delta J_t / E_x = \delta\sigma_0 / (1 - i\omega_c \tau_\varepsilon) \tag{2.41}$$

where $\delta\sigma_0$ is the contribution of this group of electrons to the zero-field conductivity, and τ_ε is their relaxation time. As in (2.20), $\delta J_t = \delta J_x + i\,\delta J_y$. Thus the angle between the current and the field E_x varies from one group of electrons to another, if τ_ε varies: not all the electrons are carrying current in the same direction. It follows at once that the resultant total current J_t is reduced, simply because $|\boldsymbol{J}_1 + \boldsymbol{J}_2| < |\boldsymbol{J}_1| + |\boldsymbol{J}_2|$ (Fig. 2.6). We can therefore expect that the model will now show a magnetoresistance effect – an increase of resistivity with $\boldsymbol{B}$.

This is confirmed by detailed calculation. When one carries out the appropriate average of (2.41), and then inverts the result to find the resistivity ρ and the Hall coefficient R_{H}, it turns out that both ρ and R_{H} depend on B. But the effect is fairly small; for $\tau \propto \varepsilon^{-1/2}$, ρ increases by a factor $32/9\pi = 1.13$ as B increases from 0 to ∞, and $-R_{\mathrm{H}}$ falls by a factor $8/3\pi = 0.85$, from $3\pi/8ne$ to $1/ne$. Far bigger variations than these are observed experimentally, in both metals and semiconductors. To understand these, and many other things, we must first look at the influence of the ionic lattice, which we have so far neglected.

3

Crystal structures and the reciprocal lattice

3.1 CRYSTAL STRUCTURES

A lump of metal does not usually look crystalline, but that is because it is almost always *poly*crystalline. Normally, when a molten metal solidifies, many tiny crystals form, and these then grow together to form a solid polycrystalline mass. By cooling slowly and carefully, one can instead grow large single crystals, and single crystals of silicon, for example, have long been the starting material for the semiconductor industry.

Conversely, by cooling extremely rapidly, at $\sim 10^6\,\mathrm{K\,s^{-1}}$, metals can be solidified into an amorphous or glassy state, in which the spatial arrangement of the atoms or ions retains the structural disorder of the liquid state, and lacks any long-range regularity. Interest in amorphous metals and semiconductors has grown rapidly since about 1970. Still more recently, in 1984, the first example was found of a new, 'quasi-crystalline' state of matter: a rapidly-cooled sample of Al_6Mn gave sharp X-ray diffraction patterns with *five*-fold symmetry. True five-fold symmetry is crystallographically impossible (since regular pentagons or icosahedra cannot be close-packed), but this material apparently counterfeits it in an ingenious way, first explored, as a mathematical possibility, in 1974.

But amorphous or quasi-crystalline samples are very much the exceptions; for the most part, metals and semiconductors are crystalline, and to understand their properties we need to consider the behaviour of conduction electrons moving through a regular crystalline array of ions. The great majority of simple metals and semiconductors crystallize in one or other of four simple structures: the body-centred cubic (bcc), face-centred cubic (fcc), hexagonal close-packed (hcp) and diamond or zinc-blende structures. These are shown in Figs 3.1 to 3.4, and we discuss them in turn.

Any regular array of points in space can be described in terms of a set of lattice vectors

$$\boldsymbol{R} = n_1\boldsymbol{a}_1 + n_2\boldsymbol{a}_2 + n_3\boldsymbol{a}_3 \tag{3.1}$$

where n_1, n_2, n_3 are integers and $\boldsymbol{a}_1, \boldsymbol{a}_2, \boldsymbol{a}_3$ are the 'primitive vectors' of the lattice. If $\boldsymbol{a}_1, \boldsymbol{a}_2, \boldsymbol{a}_3 = a\boldsymbol{i}, a\boldsymbol{j}, a\boldsymbol{k}$, vectors of length a in the x, y, z directions, (3.1) describes a simple cubic (sc) lattice. The bcc structure has an atom at each sc lattice point $\boldsymbol{R}$, and another, as shown in Fig. 3.1, at the centre of each cubic unit cell, i.e. at $\boldsymbol{R} + \boldsymbol{s}$, where $\boldsymbol{s} = \frac{1}{2}a(\boldsymbol{i} + \boldsymbol{j} + \boldsymbol{k})$. Likewise, the fcc structure has an atom at each point $\boldsymbol{R}$, and additional atoms at the centres of the cube faces, as shown in Fig. 3.2, at the points $\boldsymbol{R} + \boldsymbol{s}_1$, $\boldsymbol{R} + \boldsymbol{s}_2$, $\boldsymbol{R} + \boldsymbol{s}_3$ where

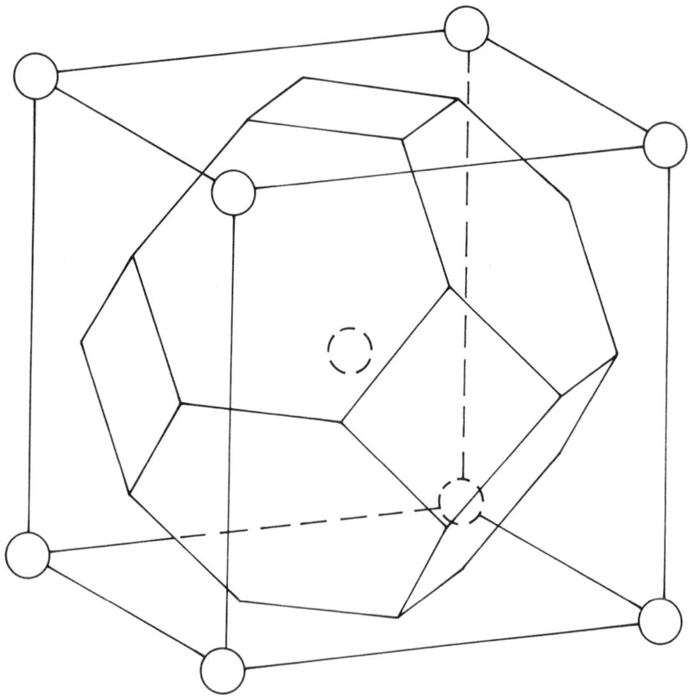

Fig. 3.1 The bcc structure, showing the Wigner–Seitz cell about the central atom.

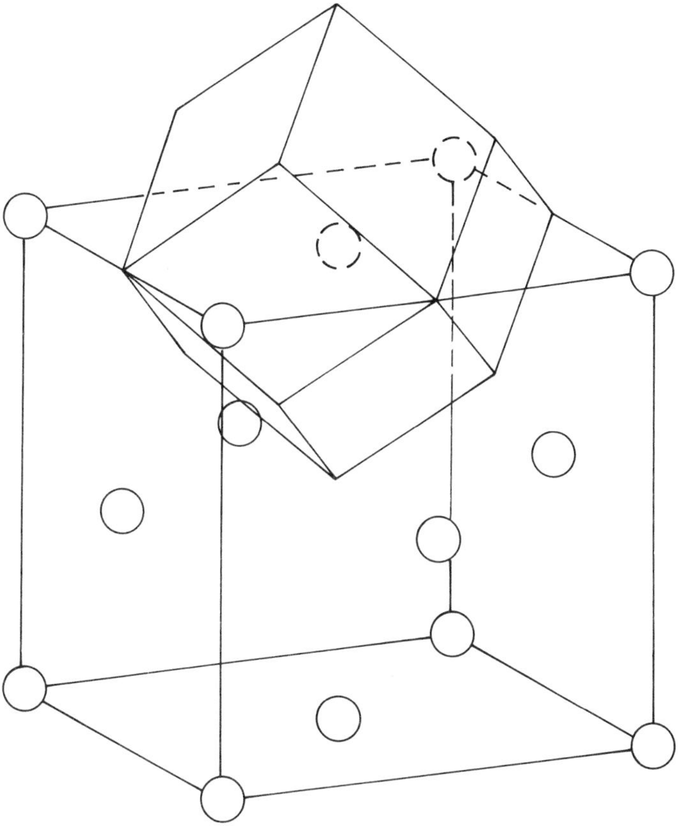

Fig. 3.2 The fcc structure, showing the Wigner–Seitz cell about the atom in the centre of the top face.

$s_1 = \frac{1}{2}a(\boldsymbol{j} + \boldsymbol{k}), s_2 = \frac{1}{2}a(\boldsymbol{k} + \boldsymbol{i}), s_3 = \frac{1}{2}a(\boldsymbol{i} + \boldsymbol{j})$. From this viewpoint, the bcc and fcc lattices are produced by adding the 'basis' $(0, \boldsymbol{s})$ or $(0, \boldsymbol{s}_1, \boldsymbol{s}_2, \boldsymbol{s}_3)$ to the set of lattice points $\boldsymbol{R}$: the bcc structure contains two atoms per unit cell, and the fcc structure contains four.

But in fact, every lattice point in the bcc lattice is in an identical environment to every other, and likewise in the fcc lattice. This is best seen by constructing the Wigner–Seitz cell around any lattice site, defined as the region nearer to that site than to any other. In the bcc lattice, the W–S cell has the shape shown in Fig. 3.1. The hexagonal faces are midway between the central site and the cube corners, and the square faces are midway between the central site and corresponding sites in adjacent cells. A little thought should convince you that every lattice site in the bcc lattice has the same environment, and therefore has a W–S cell of this shape. Likewise, Fig. 3.2 shows the fcc W–S cell around the lattice site at the centre of the top face, which has twelve diamond-shaped faces midway between it and its twelve nearest neighbours. Here again, it is not hard to see that every lattice site has the same environment, and hence has a W–S cell of the same shape.

By choosing a different set of vectors $\boldsymbol{a}_1$, $\boldsymbol{a}_2$, $\boldsymbol{a}_3$, we can in fact reach every lattice point in the bcc or fcc lattice by using (3.1) alone, without needing to introduce additional 'basis' vectors $\boldsymbol{s}$. For bcc, we can take

$$\boldsymbol{a}_1 = a\boldsymbol{i}, \quad \boldsymbol{a}_2 = a\boldsymbol{j}, \quad \boldsymbol{a}_3 = \tfrac{1}{2}a(\boldsymbol{i}+\boldsymbol{j}+\boldsymbol{k}) \qquad (3.2)$$

and for fcc

$$\boldsymbol{a}_1 = \tfrac{1}{2}a(\boldsymbol{j}+\boldsymbol{k}), \quad \boldsymbol{a}_2 = \tfrac{1}{2}a(\boldsymbol{k}+\boldsymbol{i}), \quad \boldsymbol{a}_3 = \tfrac{1}{2}a(\boldsymbol{i}+\boldsymbol{j}) \qquad (3.3)$$

You should convince yourself that (3.1) does then generate the full set of bcc or fcc lattice points. We can thus think of bcc and fcc either as sc lattices with a basis $(0, \boldsymbol{s})$ or $(0, \boldsymbol{s}_1, \boldsymbol{s}_2, \boldsymbol{s}_3)$, or as lattices with the primitive vectors (3.2) or (3.3), in which case there is no need for a basis.

The first approach makes clear the cubic symmetry of the lattice, but does not make it so clear that all lattice points are completely equivalent. The second approach does precisely the converse. In practice, the main function of the second approach is to demonstrate that the bcc and fcc lattices can indeed be described in the form (3.1), with no basis, so that they are – in crystallographers' terms – 'simple Bravais lattices'. Otherwise, one almost always thinks of bcc and fcc in terms of simple cubic axes. For example, when describing a direction in the lattice by a set of integers n_1, n_2, n_3, [111] signifies the 'body diagonal' through the centre of the cube, whether the crystal is sc, bcc or fcc.

Our other two structures – the hcp and diamond structures – cannot be represented in the form (3.1) without using a basis, whatever set of primitive vectors we choose. They are examples of a 'lattice with basis', as opposed to a 'simple Bravais lattice'. Take first the diamond lattice. Figure 3.3 marks alternate atoms differently, which makes it easier to see that this lattice consists of two interpenetrating

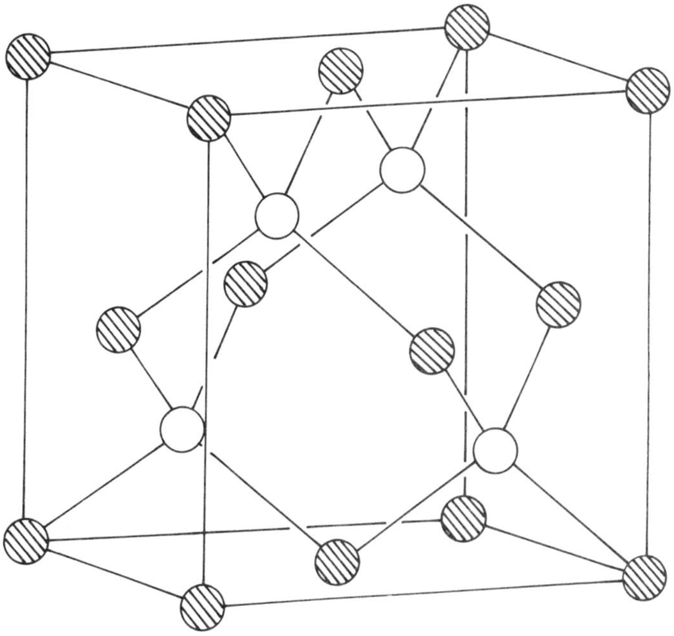

Fig. 3.3 The diamond and zinc blende structures. In diamond, and in Si and Ge, every atom is the same; in zinc blende (ZnS), alternate atoms are Zn and S, so that each Zn atom is linked to four S, and each S to four Zn. Semiconductors such as InSb and GaAs have the ZnS structure.

fcc lattices; each point of the second lattice is displaced from a point of the first by a distance $a(\boldsymbol{i}+\boldsymbol{j}+\boldsymbol{k})/4$. In GaAs or ZnS (zinc blende), we have two different atoms at the two different sites in the unit cell, to give the zinc blende structure; in Ge, Si and C (diamond), we have the same atom at both sites. In either case, the structure is basically fcc, with two atoms per unit cell if we use the cell of (3.3), or with eight atoms per cell if we use the simple cubic cell.

Either way, we have a 'lattice with basis' and not a simple Bravais lattice. This is shown, too, by the fact that not all lattice sites have identical surroundings; if we construct W–S cells around the lattice points, they will all have basically tetrahedral symmetry, but half of them will be one way up, half the other way up.

The hcp lattice, shown in Fig. 3.4, consists of two interpenetrating simple hexagonal lattices, with the layers of one half-way between those of the other. The simple hexagonal lattice can be defined by

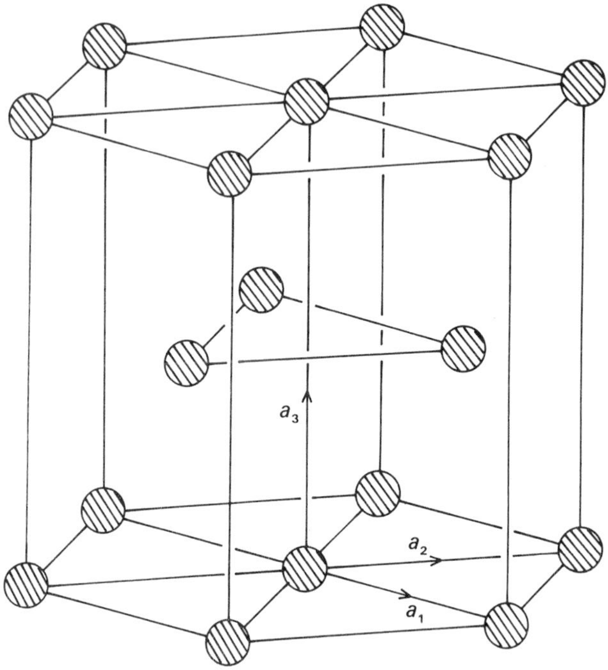

Fig. 3.4 The hcp structure.

(3.1) with $\boldsymbol{a}_1, \boldsymbol{a}_2 = a(\sqrt{3}\boldsymbol{i} \pm \boldsymbol{j})/2$, $\boldsymbol{a}_3 = c\boldsymbol{k}$, as shown. To generate the full hcp lattice we must add the set of points $\boldsymbol{R}+\boldsymbol{s}$, with $\boldsymbol{s} = a\boldsymbol{i}/\sqrt{3} + c\boldsymbol{k}/2$, and no choice of primitive vectors in (3.1) will avoid this need for a basis. Correspondingly, atoms at $\boldsymbol{R}$ and at $\boldsymbol{R}+\boldsymbol{s}$ are not identical in their surroundings, and have different W–S cells, one rotated by π about the c axis relative to the other.

There is a close relationship between the fcc and hcp structures. Both are 'close-packed', in the sense that each atom (i.e. each lattice site) is surrounded by twelve nearest neighbour sites. In the fcc structure, this is shown by the W–S cell, each of whose twelve similar faces is half-way between the central atom and one of its nearest neighbours. In the hcp structure, it is clear from Fig. 3.4 that each atom has six nearest neighbours in the same plane, three in the plane above and three in the plane below. But these will not all be at the same distance unless $|\boldsymbol{s}| = |\boldsymbol{a}_1| = a$, i.e. unless $c/a = \sqrt{(8/3)} = 1.633$. A set of spheres, stacked to form an hcp structure, will necessarily have

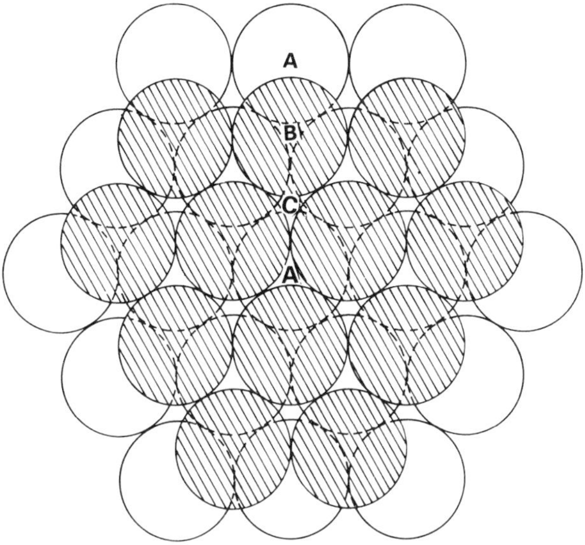

Fig. 3.5 Two successive layers of atoms in an fcc or hcp crystal. In an fcc crystal, the next layer will have atoms at points such as C; in an hcp crystal, they will be at points such as A.

this 'ideal' c/a ratio. In real hcp metals c/a varies from about 1.55 to 1.9, but such structures are still called hcp, even though not all twelve 'nearest' neighbours are equally near.

A set of spheres can equally well be stacked to form an fcc structure, in fact. As shown in Fig. 3.5, successive layers of close-packed spheres can be stacked in two ways, one of them hcp and the other fcc. From this point of view, the difference between fcc and (ideal) hcp is merely a difference in the stacking order of successive hexagonal layers.

The bcc and diamond structures are not close-packed; in a bcc crystal, each atom has only eight nearest neighbours, and in diamond only four. In fcc materials, the forces attracting the atoms to each other are non-directional, and the lowest-energy structure (and hence the stable structure) is one in which each atom is surrounded by as many others as possible. In diamond, on the other hand, the bonding forces are highly directional, and the lowest-energy state has each atom covalently bonded to four neighbours.

The crystal structures of some of the common metals and semiconductors (listed in order of their group number in the periodic table) are as follows:

bcc	: Li, Na, K, Rb, Cs; Ba; V, Nb, Ta; Cr, Mo, W; Fe.
fcc	: Cu, Ag, Au; Ca; Al; Pb; Ni, Pd, Pt.
diamond:	C, Si, Ge, α-Sn (grey tin: the semiconducting form).
hcp	: Be, Mg, Zn, Cd; Ti, Zr; Co; most rare earth metals.

Among the metals which do not crystallize in one of these four structures are Mn, Hg, In and Sn. In and β-Sn ('white' tin: the metallic form) are tetragonal. In has a slightly distorted fcc structure, in which one cube axis has been stretched by $7\frac{1}{2}\%$ relative to the other two, to become the tetragonal *c* axis. Sn has a compressed bcc structure, with the *c* axis only 55% as long as the other two axes, and two extra atoms in the unit cell. Hg has a compressed fcc structure, compressed this time along a body diagonal to make it rhombohedral, and α-Mn has a complicated cubic structure, with 58 atoms in the unit cell.

The three 'semi-metals' As, Sb and Bi (so called because although metallic they are unusually poor conductors) are all rhombohedral, though they would be simple cubic if it were not for a slight distortion along one body diagonal of the cube.

In a cubic material, the ratio $\boldsymbol{J}/\boldsymbol{E}$ of current density to electric field must by symmetry be the same whether the field is applied along the x, y or z axis. It follows by superposition that the ratio must be the same for any direction of $\boldsymbol{E}$, so that $\boldsymbol{J}=\sigma\boldsymbol{E}$ with the same σ for all $\boldsymbol{E}$ directions, and the material is said to be *isotropic*. This symmetry argument is very general, and does not depend on any detailed model of the conduction process, though Ohm's law must hold so that the principle of superposition can be used. In a material of lower symmetry, the argument can still be applied, but now it does not lead to complete isotropy. In a hexagonal or tetragonal material, for example, symmetry requires J/E to be the same for all directions of $\boldsymbol{E}$ in the 'basal plane', normal to the c axis: $\boldsymbol{J}_\perp=\sigma_\perp \boldsymbol{E}_\perp$. But there is no reason, on symmetry grounds, why the current should be the same when the field is applied *along* the c axis, and in general it will not be, so that we then have $\boldsymbol{J}_\parallel=\sigma_\parallel \boldsymbol{E}_\parallel$, with $\sigma_\parallel \neq \sigma_\perp$. In practice, $\sigma_\parallel$ and $\sigma_\perp$ do indeed differ in such materials, typically by 20% or so. In these anisotropic materials, $\boldsymbol{J}$ will not be parallel to $\boldsymbol{E}$ unless $\boldsymbol{E}$ is along the c axis or normal to it, though the angle between them will usually be small (Problem 3.2).

These anisotropies cannot be accounted for on the simple

free-electron model of Chapter 1; they arise because in real metals the Fermi surface is not simply a sphere but has a more complex shape, which reflects the crystal symmetry.

Lastly, as we saw in section 2.4, even a free-electron metal ceases to be isotropic in an applied field $\boldsymbol{B}$, and so does a cubic material, because $\boldsymbol{B}$ reduces the symmetry of the problem.

3.2 THE RECIPROCAL LATTICE

If a crystal has an atom located at each lattice point (3.1), it will show perfect periodicity: that is, the conditions at any point $\boldsymbol{r}$ in the crystal will be exactly repeated at all other points $\boldsymbol{r} + \boldsymbol{R}$ where $\boldsymbol{R}$ is any of the vectors (3.1). Thus a conduction electron moving through the crystal will have an electrostatic potential energy $V(\boldsymbol{r})$, due to its Coulomb interaction with the ions and with all the other conduction electrons, which is itself perfectly periodic:

$$V(\boldsymbol{r}) = V(\boldsymbol{r} + \boldsymbol{R}) \tag{3.4}$$

It must therefore be possible to express $V(\boldsymbol{r})$ as a three-dimensional Fourier series. To see how, consider first the equivalent one-dimensional problem: $V(x) = V(x + X)$, where $X = na$. In this case, we know that $V(x)$ can be written in the form

$$V(x) = \sum V_K \, \mathrm{e}^{iKx} \tag{3.5}$$

where the sum extends over all those values of K which satisfy $\mathrm{e}^{iKX} = 1$, so that $\mathrm{e}^{iKx} = \mathrm{e}^{iK(X+x)}$, as required. Since $X = na$, $\mathrm{e}^{iKX} = (\mathrm{e}^{iKa})^n$, and we therefore need $\mathrm{e}^{iKa} = 1$, and hence $Ka = 2\pi m$, with m an integer. These values of K, and only these values, will contribute to (3.5) terms having the required periodicity. In exactly the same way, we can write $V(\boldsymbol{r})$ in terms of its Fourier components as

$$V(\boldsymbol{r}) = \sum V_{\boldsymbol{K}} \, \mathrm{e}^{i\boldsymbol{K}\cdot\boldsymbol{r}} \tag{3.6}$$

The condition (3.4) now requires that $\mathrm{e}^{i\boldsymbol{K}\cdot\boldsymbol{R}} = 1$, or that

$$\boldsymbol{K}\cdot\boldsymbol{R} = 2\pi N \tag{3.7}$$

with N an integer, for all $\boldsymbol{R}$ defined by (3.1). The only values of $\boldsymbol{K}$

allowed in (3.6) are those that satisfy (3.7), because only then will (3.6) satisfy (3.4).

It is not too difficult to find the set of vectors $\boldsymbol{K}$ satisfying (3.7), by starting from the primitive vectors $\boldsymbol{a}_1, \boldsymbol{a}_2, \boldsymbol{a}_3$ of (3.1). Choose a vector $\boldsymbol{b}_1$ normal to $\boldsymbol{a}_2$ and $\boldsymbol{a}_3$, so that $\boldsymbol{a}_2 \cdot \boldsymbol{b}_1 = \boldsymbol{a}_3 \cdot \boldsymbol{b}_1 = 0$, and of length (in m^{-1}) such that $\boldsymbol{a}_1 \cdot \boldsymbol{b}_1 = 2\pi$. Define $\boldsymbol{b}_2$ and $\boldsymbol{b}_3$ similarly. (We can express these conditions compactly by writing

$$\boldsymbol{a}_i \cdot \boldsymbol{b}_j = 2\pi\delta_{ij} \tag{3.8}$$

where $i, j = 1, 2$ or 3 and $\delta_{ij} = 1$ if $i = j$, $\delta_{ij} = 0$ if $i \neq j$.) Then clearly any vector $\boldsymbol{K}$ defined by

$$\boldsymbol{K} = m_1\boldsymbol{b}_1 + m_2\boldsymbol{b}_2 + m_3\boldsymbol{b}_3 \tag{3.9}$$

where m_1, m_2, m_3 are integers, will satisfy (3.7), with $N = m_1 n_1 + m_2 n_2 + m_3 n_3$.

In section 1.4 we introduced the idea of '$\boldsymbol{k}$-space'. Equation (3.9) defines a set of points in $\boldsymbol{k}$-space (or 'reciprocal space') called the reciprocal lattice, just as (3.1) defines a set of points in real space called the real-space lattice. The reciprocal lattice plays a central part in any discussion of wave propagation through a crystal, whether the waves are X-rays, electrons, neutrons or phonons (sound waves). For example, Bragg reflection (or, more properly, Bragg diffraction) of a wave will occur when the wave-vectors $\boldsymbol{k}$ and $\boldsymbol{k}'$ of the incident and diffracted waves are related by $\boldsymbol{k}' = \boldsymbol{k} + \boldsymbol{K}$. But this is looking ahead; for the moment, the importance of the reciprocal lattice is that $V(\boldsymbol{r})$ can be written in terms of a set of Fourier components (3.6), where the only allowed values of $\boldsymbol{K}$ are those given by (3.9). If $V(\boldsymbol{r})$ is to be real, the coefficients $V_{\boldsymbol{K}}$ must satisfy $V_{-\boldsymbol{K}} = V^*_{\boldsymbol{K}}$. If the crystal has a centre of symmetry, so that $V(\boldsymbol{r}) = V(-\boldsymbol{r})$, this condition reduces to $V_{-\boldsymbol{K}} = V_{\boldsymbol{K}}$, since $V_{\boldsymbol{K}}$ itself will then be real.

The primitive vectors $\boldsymbol{b}_1, \boldsymbol{b}_2, \boldsymbol{b}_3$ of the reciprocal lattice are given by

$$\boldsymbol{b}_1 = 2\pi(\boldsymbol{a}_2 \times \boldsymbol{a}_3)/V_{ru}; \quad \boldsymbol{b}_2 = 2\pi(\boldsymbol{a}_3 \times \boldsymbol{a}_1)/V_{ru}; \quad \boldsymbol{b}_3 = 2\pi(\boldsymbol{a}_1 \times \boldsymbol{a}_2)/V_{ru} \tag{3.10}$$

where $V_{ru} = \boldsymbol{a}_1 \cdot (\boldsymbol{a}_2 \times \boldsymbol{a}_3) = \boldsymbol{a}_2 \cdot (\boldsymbol{a}_3 \times \boldsymbol{a}_1) = \boldsymbol{a}_3 \cdot (\boldsymbol{a}_1 \times \boldsymbol{a}_2)$ is the volume of the unit cell in real space defined by $\boldsymbol{a}_1, \boldsymbol{a}_2, \boldsymbol{a}_3$. It is easily verified that the vectors (3.10) satisfy the defining conditions (3.8). From

(3.10) we can derive an important relation between the volume of the unit cell in $\boldsymbol{k}$-space, $V_{ku} = \boldsymbol{b}_1 \cdot (\boldsymbol{b}_2 \times \boldsymbol{b}_3)$, and that in real space, V_{ru}. By writing $V_{ku} = 2\pi(\boldsymbol{a}_2 \times \boldsymbol{a}_3) \cdot (\boldsymbol{b}_2 \times \boldsymbol{b}_3)/V_{ru}$ and using a little vector algebra, we find that

$$V_{ku} = 8\pi^3/V_{ru} \tag{3.11}$$

What reciprocal lattices correspond to the four different real-space lattices that we have considered? Using (3.10) with either (3.2) or (3.3), we find a pleasing symmetry between bcc and fcc: for a bcc real-space lattice the reciprocal lattice is fcc, and conversely (problem 3.4). When the lattice has a basis, as the hcp and diamond lattices do, the reciprocal lattice is not affected by the presence of the basis, though the extra symmetry produced by the basis may cause some of the Fourier coefficients $V_{\boldsymbol{K}}$ to vanish: one says that the 'structure factor' for these values of $\boldsymbol{K}$ vanishes. Thus for diamond, which is an fcc lattice with basis, the reciprocal lattice is bcc, and for the hcp structure, the reciprocal lattice is simple hexagonal, as it would be for a simple hexagonal real-space lattice.

If a bcc or fcc lattice is regarded as an sc lattice with basis, then the sc reciprocal lattice will have $\boldsymbol{b}_1 = 2\pi\boldsymbol{i}/a$; $\boldsymbol{b}_2 = 2\pi\boldsymbol{j}/a$; $\boldsymbol{b}_3 = 2\pi\boldsymbol{k}/a$. The actual bcc or fcc reciprocal lattice is obtained from this by retaining only those lattice points (3.9) for which $m_1 + m_2 + m_3$ is even (for bcc) or for which m_1, m_2 and m_3 are either all even or all odd (for fcc). These are the only lattice points, from this viewpoint, for which the structure factor is non-zero.

4

Electrons in a periodic potential

4.1 BLOCH'S THEOREM

An electron moving through the periodic potential $V(\boldsymbol{r})$ must satisfy the Schrödinger equation

$$-(\hbar^2/2m)\nabla^2\psi + V(\boldsymbol{r})\psi = \varepsilon\psi \tag{4.1}$$

Bloch showed in 1928 that because of the periodicity of $V(\boldsymbol{r})$, the solutions $\psi(\boldsymbol{r})$ must have the form

$$\psi(\boldsymbol{r}) = \mathrm{e}^{i\boldsymbol{k}\cdot\boldsymbol{r}}u_{\boldsymbol{k}}(\boldsymbol{r}) \tag{4.2}$$

where $u_{\boldsymbol{k}}(\boldsymbol{r})$, like $V(\boldsymbol{r})$, has the periodicity of the lattice:

$$u_{\boldsymbol{k}}(\boldsymbol{r}) = u_{\boldsymbol{k}}(\boldsymbol{r}+\boldsymbol{R}) \tag{4.3}$$

for any lattice vector $\boldsymbol{R}$. Clearly any function $\psi(\boldsymbol{r})$ can formally be written as the product of a plane wave $\mathrm{e}^{i\boldsymbol{k}\cdot\boldsymbol{r}}$ and some function $u_{\boldsymbol{k}}(\boldsymbol{r})$; the essential feature of Bloch's theorem is that the function $u_{\boldsymbol{k}}(\boldsymbol{r})$ so defined has the periodicity (4.3).

We can derive (4.3) as follows. Because of the translational symmetry of the lattice, the environment around any point $\boldsymbol{r}+\boldsymbol{R}$ is identical with the environment around $\boldsymbol{r}$. We therefore expect $|\psi|$ at $\boldsymbol{r}+\boldsymbol{R}$ to be identical with $|\psi|$ at $\boldsymbol{r}$, so that $\psi(\boldsymbol{r}+\boldsymbol{R})$ can differ from $\psi(\boldsymbol{r})$ at most by a phase factor:

$$\psi(\boldsymbol{r}+\boldsymbol{R}) = \psi(\boldsymbol{r})\,\mathrm{e}^{i\theta(\boldsymbol{R})} \tag{4.4}$$

The phase angle θ will depend on $\boldsymbol{R}$, but because of translational symmetry the difference $\theta(\boldsymbol{R}_A + \boldsymbol{R}_B) - \theta(\boldsymbol{R}_A)$ between two lattice sites $\boldsymbol{R}_A$ and $\boldsymbol{R}_A + \boldsymbol{R}_B$ can depend only on their separation $\boldsymbol{R}_B$, and not on $\boldsymbol{R}_A$. This means that θ must depend *linearly* on the three integers n_1, n_2, n_3 specifying $\boldsymbol{R}$ (equation 3.1), and we can write

$$\theta = 2\pi(n_1 c_1 + n_2 c_2 + n_3 c_3)$$

where c_1, c_2, c_3 are constants for a given solution wave-function $\psi(\boldsymbol{r})$. We can write this in the equivalent form $\theta = \boldsymbol{k} \cdot \boldsymbol{R}$, where $\boldsymbol{k} = c_1 \boldsymbol{b}_1 + c_2 \boldsymbol{b}_2 + c_3 \boldsymbol{b}_3$. Equation (4.4) thus becomes

$$\psi(\boldsymbol{r} + \boldsymbol{R}) = \psi(\boldsymbol{r})\, e^{i\boldsymbol{k}\cdot\boldsymbol{R}} \tag{4.5}$$

and this result is equivalent to the periodicity condition (4.3): if we use (4.2) to write $\psi(\boldsymbol{r})$ and $\psi(\boldsymbol{r} + \boldsymbol{R})$ in terms of $u_{\boldsymbol{k}}(\boldsymbol{r})$ and $u_{\boldsymbol{k}}(\boldsymbol{r} + \boldsymbol{R})$, (4.3) follows at once.

The form of the Bloch function (4.2) is intuitively reasonable, at least for small $V(\boldsymbol{r})$. For $V(\boldsymbol{r}) = 0$, we have free electrons with $\psi = e^{i\boldsymbol{k}\cdot\boldsymbol{r}}$, and it is not surprising that a small periodic potential $V(\boldsymbol{r})$ merely produces a small periodic modulation of this plane wave. What is perhaps more surprising is that $\psi(\boldsymbol{r})$ still has the form (4.2) however strong $V(\boldsymbol{r})$ becomes.

4.2 THE BRILLOUIN ZONE

The periodic form of $V(\boldsymbol{r})$ and of $u_{\boldsymbol{k}}(\boldsymbol{r})$ has two closely-related consequences. First, it turns out that there are now many different solutions of the Schrödinger equation (4.1) for a given value of $\boldsymbol{k}$, so that we need to distinguish the different solutions by using a 'band index' n, and write

$$\psi_{n\boldsymbol{k}}(\boldsymbol{r}) = e^{i\boldsymbol{k}\cdot\boldsymbol{r}} u_{n\boldsymbol{k}}(\boldsymbol{r}) \tag{4.6}$$

though we shall often drop the subscript when no confusion arises. The energy ε of state $\psi_{n\boldsymbol{k}}(\boldsymbol{r})$ will depend on both $\boldsymbol{k}$ and n, and we write it as $\varepsilon_n(\boldsymbol{k})$.

Secondly, the form of (4.2) leads to a curious non-uniqueness in the definition of $\boldsymbol{k}$ itself. If we write $\boldsymbol{k}' = \boldsymbol{k} + \boldsymbol{K}$, where $\boldsymbol{K}$ is any reciprocal lattice vector (3.9), (4.2) becomes

$$\psi(\boldsymbol{r}) = e^{i\boldsymbol{k}'\cdot\boldsymbol{r}}[u_{\boldsymbol{k}}(\boldsymbol{r})\, e^{-i\boldsymbol{K}\cdot\boldsymbol{r}}]$$

But $e^{-i\boldsymbol{K}\cdot\boldsymbol{r}}$, like $u_{\boldsymbol{k}}(\boldsymbol{r})$ itself, has the periodicity of the lattice, so that the product [...] also has this periodicity. We thus find that the wave-function (4.2) can equally well be written in the form

$$\psi(\boldsymbol{r}) = e^{i\boldsymbol{k}'\cdot\boldsymbol{r}} u_{\boldsymbol{k}'}(\boldsymbol{r}) \tag{4.7}$$

where $u_{\boldsymbol{k}'}(\boldsymbol{r}) = u_{\boldsymbol{k}}(\boldsymbol{r})e^{-i\boldsymbol{K}\cdot\boldsymbol{r}}$ still satisfies the Bloch periodicity condition (4.3). Equations (4.2) and (4.7) are simply two different, but entirely equivalent, descriptions of the same wave-function. Consequently this wave-function can equally well be associated with the wave-vector $\boldsymbol{k}$ or with any of an infinite number of other wave-vectors $\boldsymbol{k}' = \boldsymbol{k} + \boldsymbol{K}$, repeating periodically throughout $\boldsymbol{k}$-space. Likewise, the energy $\varepsilon_n(\boldsymbol{k})$ can equally well be associated with any of the wave-vectors $\boldsymbol{k} + \boldsymbol{K}$, so that we can write

$$\varepsilon_n(\boldsymbol{k}) = \varepsilon_n(\boldsymbol{k} + \boldsymbol{K}) \tag{4.8}$$

Thus $\varepsilon_n(\boldsymbol{k})$ repeats periodically in $\boldsymbol{k}$-space in much the same way that $V(\boldsymbol{r})$ repeats periodically in real space. Although there is less physical significance in the periodicity of $\varepsilon_n(\boldsymbol{k})$, which merely reflects a periodic mathematical re-labelling of the same physical state, it is sometimes convenient to think of the energy spectrum $\varepsilon_n(\boldsymbol{k})$ as repeated periodically in this way. But if (for a given potential $V(\boldsymbol{r})$) we know the form of $\psi_{n\boldsymbol{k}}(\boldsymbol{r})$ and $\varepsilon_n(\boldsymbol{k})$ for all points $\boldsymbol{k}$ (and for all bands n) in one unit cell of the reciprocal lattice, we have all the physically significant information; in the rest of $\boldsymbol{k}$-space, the same information is merely repeated periodically.

The particular unit cell which is almost invariably chosen is the Brillouin zone (BZ), which is the exact $\boldsymbol{k}$-space analogue of the Wigner–Seitz cell in real space; it consists of all points in $\boldsymbol{k}$-space closer to one particular lattice point $\boldsymbol{K}$ than to any other lattice point. For an fcc crystal (or for a diamond structure), with a bcc reciprocal lattice, the BZ thus has the same shape as the truncated octahedron of Fig. 3.1, and for a bcc crystal, with an fcc reciprocal lattice, it has the same shape as the dodecahedron of Fig. 3.2. For an hcp crystal, the BZ is a hexagonal prism.

How many electrons can be accommodated in a given energy band? In other words, how many states are there in the BZ? For free electrons, we saw that if we took a crystal of volume V_r and imposed periodic boundary conditions to determine the allowed values of $\boldsymbol{k}$, the number of states δn_s available in volume element

$\delta^3\boldsymbol{k}$ was given by

$$\delta n_s = V_r \delta^3 \boldsymbol{k}/4\pi^3 \tag{1.23}$$

Exactly the same argument leads to exactly the same result for Bloch electrons, with wave-functions of the form (4.6). But since the BZ is of finite volume in $\boldsymbol{k}$-space, and since one BZ contains all the physically distinct values of $\boldsymbol{k}$, (1.23) sets a limit to the number of states available in one energy band. Suppose the crystal contains N unit cells, so that $V_r = NV_{ru}$. Then the number of states available in the BZ, of volume V_{ku}, is

$$n_s = NV_{ru}V_{ku}/4\pi^3$$

so that, from (3.11), we find the simple and important result

$$n_s = 2N \tag{4.9}$$

The real-space unit cell here is the cell defined by (3.2) or (3.3) for the bcc, fcc or diamond structures, or the simple hexagonal unit cell for the hcp structure. It therefore contains either one atom (bcc, fcc) or two (diamond, hcp). What (4.9) then tells us is that the BZ can accommodate, in each energy band, exactly two electrons per atom (one of spin up and one of spin down) in a bcc or fcc crystal, and exactly one electron per atom in a diamond-structure or hcp crystal.

4.3 NEARLY FREE ELECTRONS

We have seen that for electrons in a periodic potential $V(\boldsymbol{r})$ – Bloch electrons – $\varepsilon_n(\boldsymbol{k})$ repeats periodically throughout $\boldsymbol{k}$-space. But for $V(\boldsymbol{r}) \to 0$, we would expect to regain the free-electron picture; that is, we would expect to find $\varepsilon(\boldsymbol{k}) = \hbar^2k^2/2m$, with no apparent sign of periodicity. To see how these two very different pictures can be reconciled, consider a small potential $V(\boldsymbol{r})$ consisting initially of just one sinusoidal component:

$$V(\boldsymbol{r}) = 2V_1 \cos \boldsymbol{K}_1 \cdot \boldsymbol{r} = V_1(e^{i\boldsymbol{K}_1\cdot\boldsymbol{r}} + e^{-i\boldsymbol{K}_1\boldsymbol{r}}) \tag{4.10}$$

We can insert this $V(\boldsymbol{r})$ in the Schrödinger equation, and look for solutions of the Bloch-wave form (4.6), assuming that V_1 is small. In doing so, we shall learn quite a lot about the general form of the solutions, which will remain true even when $V(r)$ is not small and contains many Fourier components.

Since $u_{nk}(\boldsymbol{r})$ is periodic we can expand it, like $V(\boldsymbol{r})$ itself, in the form

$$u_{nk}(\boldsymbol{r}) = \sum c_{\boldsymbol{K}} \mathrm{e}^{-i\boldsymbol{K}\cdot\boldsymbol{r}} \tag{4.11}$$

where for later convenience we have written $c_{\boldsymbol{K}}$ as the coefficient of the $\mathrm{e}^{-i\boldsymbol{K}\cdot\boldsymbol{r}}$ term rather than the $\mathrm{e}^{i\boldsymbol{K}\cdot\boldsymbol{r}}$ term. The coefficients $c_{\boldsymbol{K}}$ will of course depend on n and $\boldsymbol{k}$, as well as on $\boldsymbol{K}$. If, as in (4.10), $V(\boldsymbol{r})$ has only two components $\pm \boldsymbol{K}_1$, it turns out that in (4.11) we need only include the components $\boldsymbol{K} = q\boldsymbol{K}_1$, with q an integer (positive or negative). We can thus write

$$u_{nk}(\boldsymbol{r}) = \sum c_q \mathrm{e}^{-iq\boldsymbol{K}_1\cdot\boldsymbol{r}} \tag{4.12}$$

summed over all integers q.

Inserting (4.10) and (4.12) into (4.1), we find

$$(\hbar^2/2m)\sum c_q(\boldsymbol{k} - q\boldsymbol{K}_1)^2 \mathrm{e}^{-iq\boldsymbol{K}_1\cdot\boldsymbol{r}} + V_1 \sum c_q [\mathrm{e}^{-i(q-1)\boldsymbol{K}_1\cdot\boldsymbol{r}} + \mathrm{e}^{-i(q+1)\boldsymbol{K}_1\cdot\boldsymbol{r}}]$$
$$= \varepsilon \sum c_q \mathrm{e}^{-iq\boldsymbol{K}_1\cdot\boldsymbol{r}} \tag{4.13}$$

where we have dropped the common factor $\mathrm{e}^{i\boldsymbol{k}\cdot\boldsymbol{r}}$ from all terms. The second term on the left can be rewritten as $V_1 \sum (c_{q+1} + c_{q-1}) \mathrm{e}^{-iq\boldsymbol{K}_1\cdot\boldsymbol{r}}$, and if we then multiply both sides of (4.13) by $\mathrm{e}^{ip\boldsymbol{K}_1\cdot\boldsymbol{r}}$ and integrate over all space, only the terms with $q = p$ survive, and we are left with the set of equations (one for each integer p):

$$(\hbar^2/2m)(\boldsymbol{k} - p\boldsymbol{K}_1)^2 c_p + V_1(c_{p+1} + c_{p-1}) = \varepsilon c_p \tag{4.14}$$

Suppose first that $V_1 = 0$. Then this equation has (for given p) two solutions; either ε satisfies

$$(\hbar^2/2m)(\boldsymbol{k} - p\boldsymbol{K}_1)^2 = \varepsilon \tag{4.15}$$

or else, necessarily, $c_p = 0$. Now if $c_p = 0$ for all p except $p = 0$, (4.12) reduces to $u_{nk}(\boldsymbol{r}) = c_0$, so that $\psi(\boldsymbol{r}) = c_0 \mathrm{e}^{i\boldsymbol{k}\cdot\boldsymbol{r}}$, precisely the free-electron form. For $p = 0$, (4.15) also reduces to precisely the free-electron form. So far so good. But what if, instead, $c_p = 0$ for all p except $p = p'$ say? Then $u_{nk}(\boldsymbol{r}) = c_{p'} \mathrm{e}^{-ip'\boldsymbol{K}_1\cdot\boldsymbol{r}}$ and

$$\psi(\boldsymbol{r}) = \mathrm{e}^{i(\boldsymbol{k} - p'\boldsymbol{K}_1)\cdot\boldsymbol{r}} = \mathrm{e}^{i\boldsymbol{k}'\cdot\boldsymbol{r}} \quad \text{say} \tag{4.16}$$

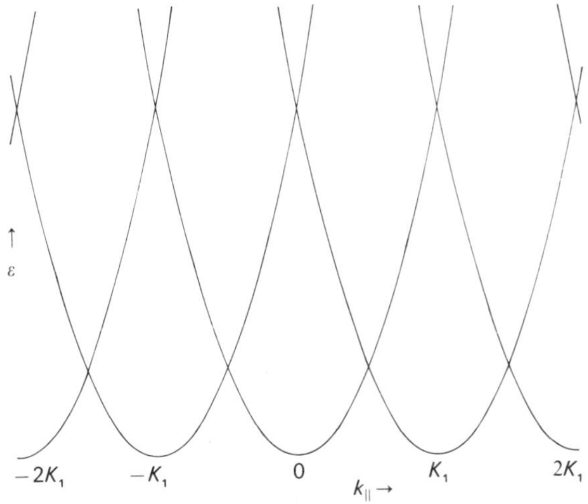

Fig. 4.1 The set of parabolas generated by (4.15), for $\boldsymbol{k} \| \boldsymbol{K}_1$.

– a plane wave whose 'real' wave-vector is $\boldsymbol{k}' = \boldsymbol{k} - p'\boldsymbol{K}_1$, not $\boldsymbol{k}$, and whose energy is correspondingly $(\hbar^2/2m)(\boldsymbol{k} - p'\boldsymbol{K}_1)^2$, precisely as given by (4.15). If we plot ε against $\boldsymbol{k}$, for $\boldsymbol{k} \| \boldsymbol{K}_1$, equation (4.15) generates a whole set of parabolas as shown in Fig. 4.1, one for each value of p. Each parabola merely repeats the same information, but with $\boldsymbol{k}$ re-labelled as $\boldsymbol{k} - p\boldsymbol{K}_1$, and $u_{n\boldsymbol{k}}(\boldsymbol{r})$ adjusted accordingly. What has happened is that the mathematics has generated for us precisely the periodicity of $\varepsilon(\boldsymbol{k})$ predicted in section (4.2). This effect of the periodic $V(\boldsymbol{r})$ remains, like the grin of the Cheshire Cat, even after we have set $V_1 = 0$.

Not only is $\varepsilon(\boldsymbol{k})$ now periodic; at the same time, as Fig. 4.1 makes clear, $\varepsilon(\boldsymbol{k})$ becomes many-valued. For given $\boldsymbol{k}$, there are many different values of $\varepsilon(\boldsymbol{k})$ corresponding to different parabolas, which we can label $\varepsilon_1(\boldsymbol{k}), \varepsilon_2(\boldsymbol{k}), \ldots$. in order of increasing energy.

Figure 4.1 represents a one-dimensional plot of $\varepsilon(\boldsymbol{k})$ for $\boldsymbol{k} \| \boldsymbol{K}_1$. In three dimensions, we can visualize (4.15) as defining, for given ε, a set of spheres in $\boldsymbol{k}$-space, each of radius $(2m\varepsilon/\hbar^2)^{1/2}$, with one sphere centred on each lattice point $p\boldsymbol{K}_1$. Wherever two of these spheres intersect (or wherever two parabolas cross in Fig. 4.1) it means that (4.15) is satisfied for the same $\varepsilon(\boldsymbol{k})$ for two different values of p, p_1 and p_2 say. We must then have

$$(\boldsymbol{k} - p_1\boldsymbol{K}_1)^2 = (\boldsymbol{k} - p_2\boldsymbol{K}_1)^2$$

which, on multiplying out, reduces to

$$\boldsymbol{k}\cdot\boldsymbol{K}_1 = \tfrac{1}{2}(p_1 + p_2)\boldsymbol{K}_1^2 \tag{4.17}$$

If we write $\boldsymbol{k} = \boldsymbol{k}_\parallel + \boldsymbol{k}_\perp$, with $\boldsymbol{k}_\parallel \parallel \boldsymbol{K}_1$ and $\boldsymbol{k}_\perp \perp \boldsymbol{K}_1$, this equation becomes $\boldsymbol{k}_\parallel = \frac{1}{2}(p_1 + p_2)\boldsymbol{K}_1$, and defines a set of planes in $\boldsymbol{k}$-space, normal to the direction of $\boldsymbol{K}_1$ and spaced at intervals $\frac{1}{2}\boldsymbol{K}_1$. Unless $\boldsymbol{k}$ lies on one of these planes, (4.15) can only be satisfied, for given ε, by at most one value of p, and therefore at most one coefficient c_p can be non-zero. But on the plane (4.17), both c_{p_1} and c_{p_2} can be non-zero simultaneously.

This result is immediately relevant when we move away from $V_1 = 0$, and consider a small but finite V_1. Suppose we are looking for a solution of the set of equations (4.14) for $\boldsymbol{k}$ close to the plane $\boldsymbol{k}_\parallel = \frac{1}{2}\boldsymbol{K}_1$, corresponding to $p_1 = 0$, $p_2 = 1$. Then we expect to find a solution in which c_0 and c_1 are both appreciable, but all other coefficients are negligibly small. The set (4.14) then reduces to just two equations:

$$\begin{aligned} [\varepsilon - \varepsilon_0(\boldsymbol{k})]c_0 &= V_1 c_1 \\ [\varepsilon - \varepsilon_1(\boldsymbol{k})]c_1 &= V_1 c_0 \end{aligned} \tag{4.18}$$

where $\varepsilon_0(\boldsymbol{k}) = \hbar^2 k^2/2m$ and $\varepsilon_1(\boldsymbol{k}) = \hbar^2(\boldsymbol{k} - \boldsymbol{K}_1)^2/2m$. From the product of these two equations, we have

$$[\varepsilon - \varepsilon_0(\boldsymbol{k})][\varepsilon - \varepsilon_1(\boldsymbol{k})] = V_1^2 \tag{4.19}$$

For $V_1 = 0$, this has the two solutions $\varepsilon = \varepsilon_0(\boldsymbol{k})$ and $\varepsilon = \varepsilon_1(\boldsymbol{k})$ as before, crossing at $\boldsymbol{k}_\parallel = \frac{1}{2}\boldsymbol{K}_1$ (Fig. 4.1). But the effect of V_1 is to *perturb apart* these two solutions, so that instead of crossing at $\boldsymbol{k}_\parallel = \frac{1}{2}\boldsymbol{K}_1$ they become two separate branches, as shown in Fig. 4.2. On the plane $\boldsymbol{k}_\parallel = \frac{1}{2}\boldsymbol{K}_1$, we have $\varepsilon_0(\boldsymbol{k}) = \varepsilon_1(\boldsymbol{k})$, $= \varepsilon_{\mathrm{B}}$ say, and (4.19) shows that the two solutions are then given by $(\varepsilon - \varepsilon_{\mathrm{B}})^2 = V_1^2$, i.e. by

$$\varepsilon = \varepsilon_{\mathrm{B}} \pm V_1 \tag{4.20}$$

The energy gap between the two bands at this point is thus $2V_1$, just equal to the amplitude of the periodic potential (4.10). From (4.18), it follows that we then have $c_1 = \pm c_0$, so that the solution wave-functions have the form

$$\psi_{\boldsymbol{k}}(\boldsymbol{r}) = c_0 \mathrm{e}^{i\boldsymbol{k}_\perp\cdot\boldsymbol{r}}(\mathrm{e}^{i\boldsymbol{K}_1\cdot\boldsymbol{r}/2} \pm \mathrm{e}^{-i\boldsymbol{K}_1\cdot\boldsymbol{r}/2}) \tag{4.21}$$

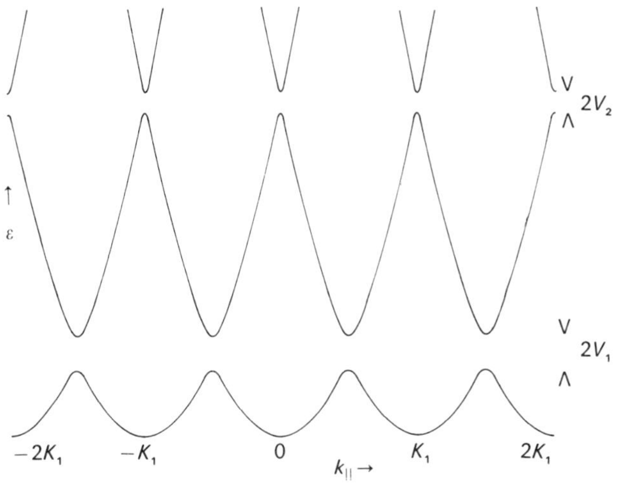

Fig. 4.2 The parabolas of Fig. 4.1, perturbed apart by a periodic potential $V(r)$.

In other words, $\psi_k(\boldsymbol{r})$ is now a standing wave varying as $\cos\frac{1}{2}\boldsymbol{K}_1\cdot\boldsymbol{r}$ or $\sin\frac{1}{2}\boldsymbol{K}_1\cdot\boldsymbol{r}$ in the $\boldsymbol{K}_1$ direction. In physical terms, we can say that when $\boldsymbol{k}_\parallel=\frac{1}{2}\boldsymbol{K}_1$, the wave $e^{i\boldsymbol{k}\cdot\boldsymbol{r}}$ is Bragg diffracted by the potential $V(\boldsymbol{r})$, and sets up a diffracted wave of equal amplitude with a wave-vector $\boldsymbol{k}'=\boldsymbol{k}-\boldsymbol{K}_1$. We can see from (4.21) how the energy gap of $2V_1$ arises. For one of the functions (4.21), $|\psi|^2=4c_0^2\cos^2\frac{1}{2}\boldsymbol{K}_1\cdot\boldsymbol{r}=2c_0^2(1+\cos\boldsymbol{K}_1\cdot\boldsymbol{r})$, and for the other, $|\psi|^2=4c_0^2\sin^2\frac{1}{2}\boldsymbol{K}_1\cdot\boldsymbol{r}=2c_0^2(1-\cos\boldsymbol{K}_1\cdot\boldsymbol{r})$. Thus for one of these functions, $|\psi|^2$ is largest where $V(\boldsymbol{r})$ is large and positive, and for the other it is largest where $V(\boldsymbol{r})$ is large and negative; the mean potential energies of these two states differ by just $2V_1$.

As $\boldsymbol{k}_\parallel$ moves away from $\frac{1}{2}\boldsymbol{K}_1$, the energies of the two bands tend rapidly back towards the free-electron values given by (4.15) (problem 4.1), and correspondingly one of the two coefficients c_0, c_1 becomes much smaller than the other. In the lower band, we have $c_0\gg c_1$ for $\boldsymbol{k}_\parallel<\frac{1}{2}\boldsymbol{K}_1$, and $c_1\gg c_0$ for $\boldsymbol{k}_\parallel>\frac{1}{2}\boldsymbol{K}_1$. As $\boldsymbol{k}_\parallel$ continues to increase and approaches $3\boldsymbol{K}_1/2$, the coefficient c_2 begins to grow, until at $\boldsymbol{k}_\parallel=3\boldsymbol{K}_1/2$, $c_2=\pm c_1$ and the whole process then repeats itself. Indeed it repeats itself whenever $\boldsymbol{k}_\parallel=(n+\frac{1}{2})\boldsymbol{K}_1$, as shown in Fig. 4.2. Note that Fig. 4.2 shows $\varepsilon(\boldsymbol{k})$ for $\boldsymbol{k}=\boldsymbol{k}_\parallel$, i.e. for $\boldsymbol{k}_\perp=0$. If $\boldsymbol{k}_\perp$ is non-zero, it merely adds the constant amount $\hbar^2k_\perp^2/2m$ to all energies and thus displaces all the curves in Fig. 4.2 upwards by that amount.

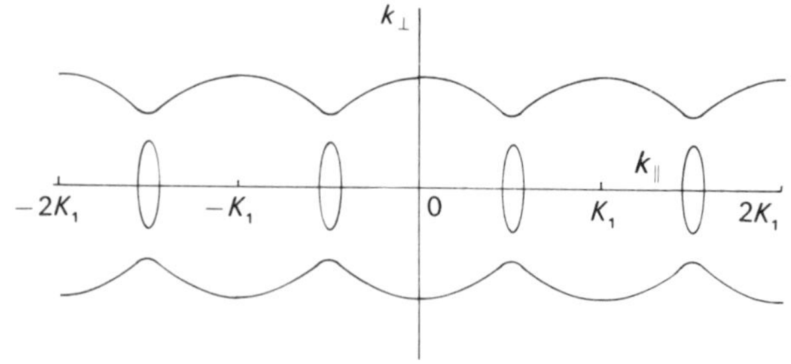

Fig. 4.3 Contours of constant energy $\varepsilon(\boldsymbol{k})$ produced by V_1.

There are three standard ways of presenting $\varepsilon(\boldsymbol{k})$ data. One is to plot ε against $\boldsymbol{k}$ along some specified *line* in $\boldsymbol{k}$-space, as in Fig. 4.2. Another is to plot a set of contours of constant energy on some specified *plane* in $\boldsymbol{k}$-space. The third is to draw a picture of some specified constant-energy *surface* (usually, for metals, the Fermi surface $\varepsilon(\boldsymbol{k}) = \varepsilon_F$) in three-dimensional $\boldsymbol{k}$-space. Figure 4.3 shows the two-dimensional equivalent of Fig. 4.2 – contours of constant energy on a plane in $\boldsymbol{k}$-space, for a particular energy which (for $\boldsymbol{k}_\perp = 0$) lies just above the gap between the first and second bands (see problem 4.2).

In this simple example, the contours will look the same whatever direction in $\boldsymbol{k}$-space (normal to $\boldsymbol{k}_\parallel$) we choose as the $\boldsymbol{k}_\perp$ axis, so that it is easy to visualize the corresponding three-dimensional picture: if Fig. 4.3 is rotated about the $\boldsymbol{k}_\parallel$ axis, it will produce a 'corrugated cylinder' running along that axis, with a 'lens' at each waist of the cylinder. The corrugated cylinder states are in the lower band, and the lens states in the upper band. We reach exactly the same picture if we start by thinking, for $V_1 = 0$, of a set of free-electron spheres in $\boldsymbol{k}$-space, one centred on each point $p\boldsymbol{K}_1$. These spheres will intersect on the planes (4.17), and where they intersect the potential V_1 perturbs the two states apart, to re-map the set of intersecting spheres into the corrugated cylinder and the set of lenses.

Having discussed at some length the effect of the very simple potential (4.10), we can now see fairly easily what the effect of more complicated potentials will be. Suppose first that $V(\boldsymbol{r})$ still has the basic periodicity described by the wave-vector $\boldsymbol{K}_1$, but now contains higher harmonics, so that

$$V(\boldsymbol{r}) = \sum V_h \mathrm{e}^{-ih\boldsymbol{K}_1\cdot\boldsymbol{r}}$$

summed over the integers h. Then (4.14) is replaced by

$$(\hbar^2/2m)(\boldsymbol{k}-p\boldsymbol{K}_1)^2 c_p + \sum V_h c_{p-h} = \varepsilon c_p \tag{4.22}$$

The effect of V_h here is to couple the equations involving c_p and $c_{p\pm h}$. V_h hence couples, and perturbs apart, the parabolas (or in three dimensions the spheres) centred on $p\boldsymbol{K}_1$ and on $(p \pm h)\boldsymbol{K}_1$, just as V_1 coupled c_p and $c_{p\pm 1}$ and hence coupled adjacent parabolas in (4.14). In Fig. 4.2 we have in fact shown the effect of V_2 as well as V_1; the upper crossing-points in Fig. 4.1 would not be perturbed apart by V_1, but they are by V_2.

Lastly, suppose that $V(\boldsymbol{r})$ has the general form

$$V(\boldsymbol{r}) = \sum V_{\boldsymbol{G}} \mathrm{e}^{-i\boldsymbol{G}\cdot\boldsymbol{r}} \tag{4.23}$$

and that, correspondingly,

$$u_{n\boldsymbol{k}}(\boldsymbol{r}) = \sum c_{\boldsymbol{K}} \mathrm{e}^{-i\boldsymbol{K}\cdot\boldsymbol{r}}. \tag{4.11}$$

Both these sums extend over all reciprocal lattice vectors, but we have labelled them $\boldsymbol{G}$ in one case and $\boldsymbol{K}$ in the other in order to distinguish them. We then find, as the full generalization of (4.14),

$$(\hbar^2/(2m)(\boldsymbol{k}-\boldsymbol{K})^2 c_{\boldsymbol{K}} + \sum_{\boldsymbol{G}} V_{\boldsymbol{G}} c_{\boldsymbol{K}-\boldsymbol{G}} = \varepsilon c_{\boldsymbol{K}} \tag{4.24}$$

– a coupled set of equations, one for each $\boldsymbol{K}$, which must be solved simultaneously to find $\varepsilon(\boldsymbol{k})$ and the coefficients $c_{\boldsymbol{K}}$.

If we assume, as before, that all the coefficients $V_{\boldsymbol{G}}$ are small – if, that is, we continue to make the 'nearly-free-electron' (NFE) approximation – the solutions follow the same form as before; the free-electron solutions $\varepsilon = (\hbar^2/2m)(\boldsymbol{k}-\boldsymbol{K})^2$ are perturbed apart by one of the Fourier coefficients $V_{\boldsymbol{G}}$ wherever two of the free-electron spheres, centred on different lattice points $\boldsymbol{K}$ and $\boldsymbol{K}-\boldsymbol{G}$, intersect. The perturbation again causes the kind of re-mapping shown in Figs 4.2 and 4.3: where the two surfaces would intersect, they split apart instead.

We can thus construct the NFE energy surfaces by drawing spheres of equal radius centred on each lattice point $\boldsymbol{K}$ – that is, at the centre of each Brillouin zone – and rejoining them in this way where they overlap, to form a set of non-intersecting surfaces. This sounds simple enough in principle, but the resultant energy surfaces

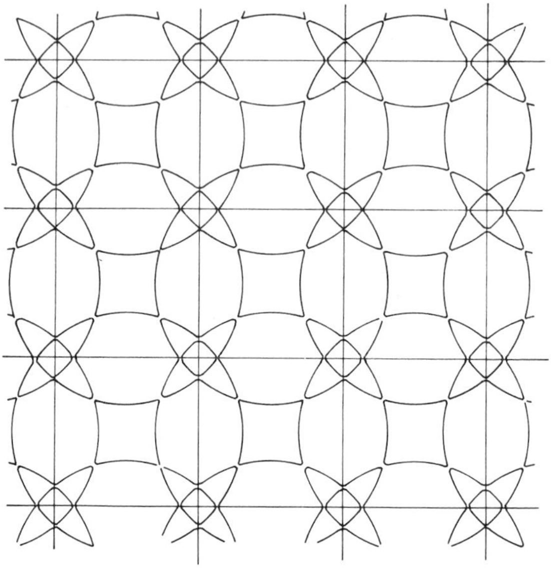

Fig. 4.4 Contours of constant energy for a simple cubic crystal, in the periodic zone scheme.

become quite complicated in practice, even for a simple cubic crystal, and even if we look at just a two-dimensional section in $\boldsymbol{k}$-space, as in Fig. 4.4. Each reciprocal lattice point is then at the centre of a cube-shaped BZ, shown as a square in this section. If the BZ has sides of length b, the biggest sphere that could be drawn around each lattice point without the spheres touching would have radius $\frac{1}{2}b$ and volume $\pi b^3/6 = 0.52b^3$. Since the volume of the whole BZ is b^3, this sphere could hold 1.04 electrons per atom, from (4.9). If our simple cubic crystal is a metal, and if this hypothetical metal contains one conduction electron per atom, the FS on this NFE approximation will be simply a sphere, because a sphere of the required volume can just fit inside the BZ without overlap.

Any larger spheres will overlap, and the circles in Fig. 4.4 represent sections through spheres of radius $0.782b$ and of volume $2b^3$, corresponding to four conduction electrons per atom. The circles have been dissected and re-assembled where they intersect, in accordance with our prescription, and the effect has been to break up the FS (or this cross-section through it) into three distinct parts, one around the zone centre and the other two at the corners (and all, of course,

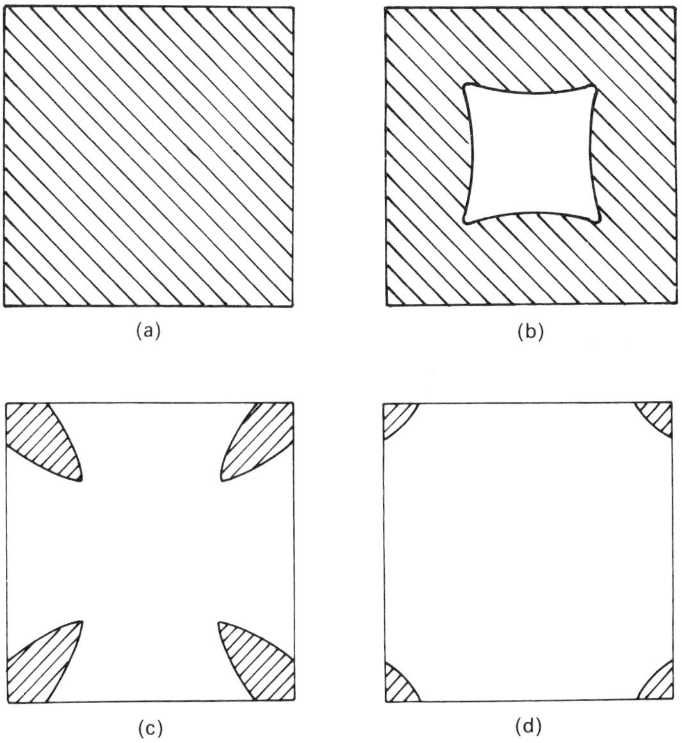

Fig. 4.5 The contours of Fig. 4.4, plotted in the reduced zone scheme. Figures (a) to (d) show bands 1 to 4 respectively. Band 1 is completely full.

repeating periodically throughout $\boldsymbol{k}$-space). Each part belongs to a different energy band, and these bands are shown separately in Fig. 4.5, in the 'reduced' zone scheme, which shows just the contents of a single BZ. This contains all the physically relevant information, although it breaks up the surfaces in bands 3 and 4 into four pieces, whereas it is clear from Fig. 4.4 that these in fact join up to form single surfaces.

In re-assembling the intersecting spheres in Fig. 4.4, and in deriving Fig. 4.5, it is useful to remember that the re-assembled surfaces must consist of segments which are all either convex outwards, as in bands 3 and 4, or concave outwards, as in band 2; otherwise we should have an impossible situation with the lower-energy states outside the surface at some points and inside it at others. Note also that the band number of a given filled region is $n + 1$, where n is the number

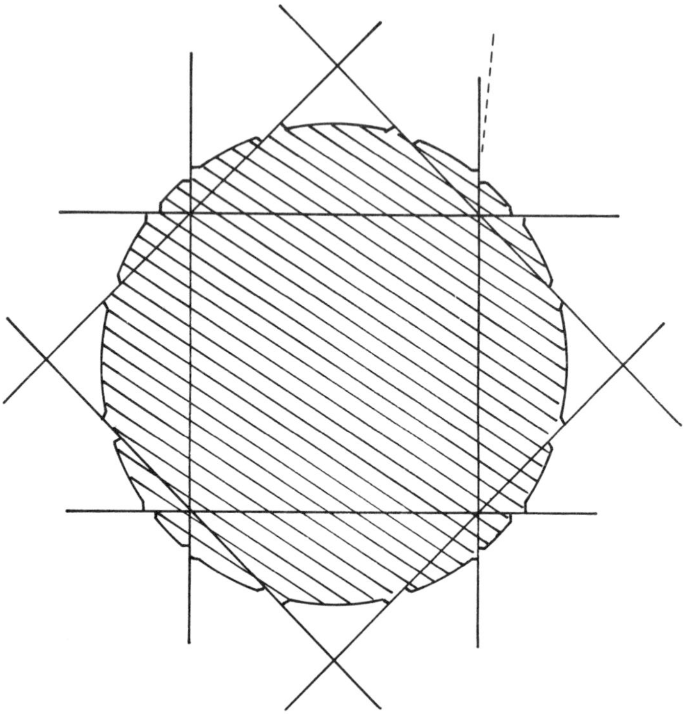

Fig. 4.6 The contours of Fig. 4.4, plotted in the extended zone scheme.

of different free-electron surfaces we pass through, for example in Fig. 4.4, to reach that region from the origin.

The first (lowest) band in Fig. 4.5 is completely full, at least in this cross-section, because the free-electron sphere is so much larger than the BZ that it extends beyond it in all directions; in other words, all states in the first band have energies less than ε_F. In the second band, the energy is higher at the centre of the zone than at the sides (just as in the second band in Fig. 4.2), and in fact $\varepsilon > \varepsilon_F$ at the centre of the zone, so that the states here are empty. In bands 3 and 4, only the states at the zone corners are filled, because only they have $\varepsilon < \varepsilon_F$.

Another way of displaying these results is to use the 'extended' zone scheme, as shown in Fig. 4.6, in which one plots only the fragments of a single free-electron sphere. This again contains all the physically relevant information, but in a much less convenient form, and is seldom met with outside introductory textbooks. It is easily seen that Fig. 4.6 contains the same information as Fig. 4.5, but with various bits of the FS displaced through a reciprocal lattice vector to another equivalent position. It follows that the total volume occupied by the filled states in $\boldsymbol{k}$-space is the same in Figs 4.5 and 4.6, as of course it must be. In the NFE approximation, it is simply the volume of the corresponding free-electron sphere.

All this would hardly be worth going through in this much detail, but for one surprising fact – the NFE approximation describes the FS of many real metals remarkably well, as we shall see in Chapter 5.

5

Electronic band structures

5.1 THE BAND STRUCTURE OF REAL METALS

In deriving (4.24), we did not have to assume that $V(\boldsymbol{r})$ was small, and in principle this set of equations enables us to find the energies $\varepsilon_n(\boldsymbol{k})$ and the wave-functions $\psi_{n\boldsymbol{k}}(\boldsymbol{r})$ for an electron moving in any periodic potential $V(\boldsymbol{r})$. But in practice, the equations can be solved at all easily only if most of the coefficients $V_{\boldsymbol{G}}$ are small (compared with ε_F, say). In the NFE approximation we assume that they are *all* small, so that the solution wave-functions contain only a few non-zero Fourier coefficients $c_{\boldsymbol{K}}$. In reality, this will not be so; $V(\boldsymbol{r})$ varies rapidly with $\boldsymbol{r}$, as shown in Fig. 5.1, and consequently the coefficients $V_{\boldsymbol{G}}$ are large, and only decrease in magnitude slowly as $|\boldsymbol{G}|$ increases. Equation (4.24) then becomes a formidably large set of coupled equations. The solution wave-functions contain many Fourier components and, within the ion core at least, they look nothing like plane waves. Physically, this is just what we should expect: within the ion cores, $\psi(\boldsymbol{r})$ must look something like an atomic wave-function, and such a function will certainly need many Fourier components to represent it adequately.

This suggests thaat a better approach for real metals may be actually to write $\psi(\boldsymbol{r})$ in terms of atomic wave-functions; if $\phi_n(\boldsymbol{r})$ is a possible wave-function for a valence electron in an isolated atom located at the origin, so that $\phi_n(\boldsymbol{r}-\boldsymbol{R})$ is the corresponding wave-function for an isolated atom located at lattice point $\boldsymbol{R}$, we can try taking

$$\psi^t_{n\boldsymbol{k}}(\boldsymbol{r}) = \sum e^{i\boldsymbol{k}\cdot\boldsymbol{R}}\phi_n(\boldsymbol{r}-\boldsymbol{R}) \tag{5.1}$$

summed over all lattice points $\boldsymbol{R}$, as an approximation to $\psi_{n\boldsymbol{k}}(\boldsymbol{r})$.

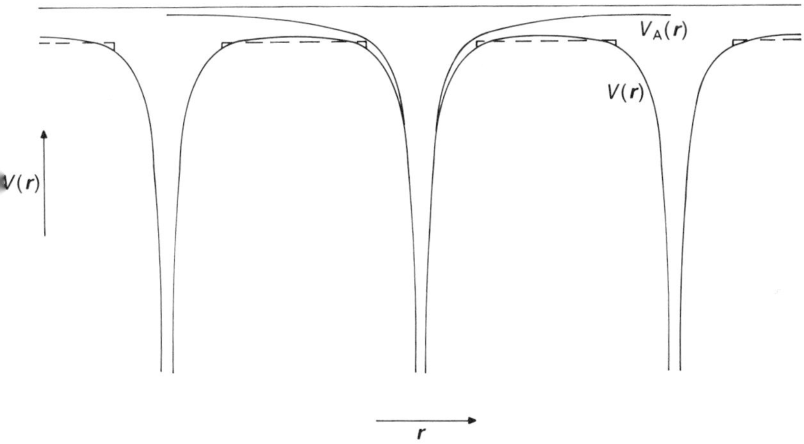

Fig. 5.1 The potential energy $V(\boldsymbol{r})$ (shown here along a line going through the atomic nuclei) is the superposition of contributions $V_A(\boldsymbol{r})$ from each ion. The broken line shows the 'muffin-tin' approximation to $V(\boldsymbol{r})$.

(There will be many different valence-electron wave-functions $\phi_n(\boldsymbol{r})$ which satisfy the Schrödinger equation for the isolated atom. We assume that these have been labelled by different subscripts m, and that the Bloch wave for band n is approximated by the set of wave-functions $\phi_n(\boldsymbol{r})$.)

The function (5.1) clearly satisfies the Bloch condition (4.5), since

$$\psi^t_{nk}(\boldsymbol{r}+\boldsymbol{R}_A)=\sum e^{i\boldsymbol{k}\cdot\boldsymbol{R}}\phi_n(\boldsymbol{r}+\boldsymbol{R}_A-\boldsymbol{R})=e^{i\boldsymbol{k}\cdot\boldsymbol{R}_A}\sum e^{i\boldsymbol{k}\cdot\boldsymbol{R}'}\phi_n(\boldsymbol{r}-\boldsymbol{R}')$$

(where $\boldsymbol{R}'=\boldsymbol{R}-\boldsymbol{R}_A$)

$$=e^{i\boldsymbol{k}\cdot\boldsymbol{R}_A}\psi^t_{nk}(\boldsymbol{r}),$$

as required.

This approach is called the tight-binding approximation (TBA), because it is useful only when $\phi_n(\boldsymbol{r}-\boldsymbol{R}_i)$ is rather closely localized around the lattice point $\boldsymbol{R}_i$. $\phi_n(\boldsymbol{r}-\boldsymbol{R}_i)$ is the solution of the Schrödinger equation for an electron moving in an isolated atom, with potential energy $V_A(\boldsymbol{r}-\boldsymbol{R}_i)$ say. Only for $\boldsymbol{r}$ close to $\boldsymbol{R}_i$ is the actual potential $V(\boldsymbol{r})$ close to $V_A(\boldsymbol{r}-\boldsymbol{R}_i)$ (cf. Fig. 5.1), so that only if $\phi_n(\boldsymbol{r}-\boldsymbol{R}_i)$ is localized in this region will (5.1) be a reasonable approximation to the true Bloch function.

Because of this, (5.1) is of limited usefulness in practice. One can try to do better by replacing the single atomic wave-function $\phi_n(\boldsymbol{r}-\boldsymbol{R})$ in (5.1) by a 'linear combination of atomic orbitals', $\sum c_m\phi_m(\boldsymbol{r}-\boldsymbol{R})$, but this LCAO method still suffers from much the same limitations. It can in fact be shown that the Bloch function $\psi_{n\boldsymbol{k}}(\boldsymbol{r})$ can be written *exactly* in the form

$$\psi_{n\boldsymbol{k}}(\boldsymbol{r})=\sum \mathrm{e}^{i\boldsymbol{k}\cdot\boldsymbol{R}}\phi_{n\mathrm{W}}(\boldsymbol{r}-\boldsymbol{R}) \tag{5.2}$$

where $\phi_{n\mathrm{W}}(\boldsymbol{r}-\boldsymbol{R})$ is called the 'Wannier function' for band n. For a tightly-bound band, $\phi_{n\mathrm{W}}(\boldsymbol{r})$ is localized closely around $\boldsymbol{r}=0$, but for bands of higher energy, less tightly bound, $\phi_{n\mathrm{W}}(\boldsymbol{r})$ may be very much less well localized. Unfortunately, (5.2) does not help us to evaluate $\psi_{n\boldsymbol{k}}(\boldsymbol{r})$ and $\varepsilon_n(\boldsymbol{k})$ in practice, because $\phi_{n\mathrm{W}}(\boldsymbol{r})$ is even more difficult to evaluate than $\psi_{n\boldsymbol{k}}(\boldsymbol{r})$ itself.

Thus neither the Fourier solution (4.24) nor the TBA solution (5.1) is powerful enough to enable the band structure – the form of $\varepsilon_n(\boldsymbol{k})$ – to be worked out accurately for a realistic potential $V(\boldsymbol{r})$. The two approaches are in effect complementary, and they fail in complementary ways: the Fourier method cannot handle the rapid variation of $V(\boldsymbol{r})$ and $\psi_{n\boldsymbol{k}}(\boldsymbol{r})$ within the ion cores, and the TBA method cannot handle the variations outside the ion cores.

These approaches, developed by Bloch and Brillouin between 1928 and 1930, were (and are) very helpful in giving a qualitative idea of the form of $\psi_{n\boldsymbol{k}}(\boldsymbol{r})$ and $\varepsilon_n(\boldsymbol{k})$, but since then a number of much more powerful techniques have been developed for actually calculating the band-structure accurately for a given $V(\boldsymbol{r})$. Three have proved particularly useful: the 'augmented plane wave' (APW) method of Slater (1937), the 'orthogonalized plane wave' (OPW) method of Herring (1940) and Green's function or KKR method of Korringa (1947) and Kohn and Rostoker (1954).

In the APW and KKR methods, the potential $V(\boldsymbol{r})$ is approximated by a 'muffin-tin' potential, spherically symmetric in a spherical region about each lattice point, and constant in the regions between the spheres (Fig. 5.1). An 'augmented' plane wave consists of a plane wave $\mathrm{e}^{i\boldsymbol{k}\cdot\boldsymbol{r}}$ in the region between the spheres, and a linear combination of atomic-like wave-functions satisfying the spherically symmetric potential within each sphere; the linear combination is chosen to match on to the plane wave at the surface of the sphere, so that the resultant APW, $\phi_{\mathrm{APW},\boldsymbol{k}}(\boldsymbol{r})$, is continuous across this surface. With this composite wave-function, the APW method combines the advantages, and avoids the disadvantages, of the Fourier approach and the TBA approach.

The Bloch wave $\psi_{nk}(\boldsymbol{r})$ is now written as a sum of APWs:

$$\psi_{nk}(\boldsymbol{r}) = \sum c_{\boldsymbol{K}} \phi_{\mathrm{APW},\boldsymbol{k}-\boldsymbol{K}}(\boldsymbol{r}),$$

and the coefficients $c_{\boldsymbol{K}}$ are chosen by a variational method to give the best solution. The equations determining these coefficients turn out to be almost identical in form to (4.24), but the coefficients $V_{\boldsymbol{G}}$ are no longer just the Fourier coefficients of $V(\boldsymbol{r})$; they are much more complicated objects, and depend on $\boldsymbol{k}$ and $\boldsymbol{K}$ as well as on $\boldsymbol{G}$. Nevertheless, relatively few of these coefficients are large enough to be important, so that the coupled set of equations is of manageable size, and can readily be solved on a computer.

The KKR method looks, on the face of it, very different from the APW method in its mathematical approach. It starts from an integral form of the Schrödinger equation, in which the Bloch wave $\psi_{nk}(\boldsymbol{r})$ at point $\boldsymbol{r}$ is regarded as made up of contributions from the wave $\psi_{nk}(\boldsymbol{r}')$ at other points $\boldsymbol{r}'$, scattered towards $\boldsymbol{r}$ by the potential $V(\boldsymbol{r}')$. For a muffin-tin potential we can put $V(\boldsymbol{r}) = 0$ outside the spherical ion-core regions, and the problem can then eventually be reduced to the solution of a set of coupled equations which are again of the form (4.24); the coefficients $V_{\boldsymbol{G}}$ are again complicated functions of $\boldsymbol{k}$ and $\boldsymbol{K}$ as well as $\boldsymbol{G}$. They are not identical with the APW coefficients $V_{\boldsymbol{G}}$; nevertheless, it can be shown that the two methods should give identical solutions if no approximations are made in solving the coupled sets of equations. In practice, the two methods do indeed give closely similar results, and so does the OPW method, when it is applicable.

The OPW method starts from the fact that any two non-degenerate solutions of a given Schrödinger equation must be orthogonal to each other: $\int \psi_1^* \psi_2 \, \mathrm{d}^3\boldsymbol{r} = 0$. Now although (5.1) may not give a good description of a valence wave-function, it should give a much better description of the tightly-bound core states, if the states $\phi_n(\boldsymbol{r} - \boldsymbol{R})$ are chosen to be atomic core wave-functions. The conduction electron wave-functions should therefore be orthogonal to these core states. A simple plane wave $\mathrm{e}^{i\boldsymbol{k}\cdot\boldsymbol{r}}$ will certainly not be orthogonal to them, but it can be made so by adding to it appropriate amounts of the core functions ψ^t_{mk}, to form an 'orthogonalized plane wave':

$$\phi_{\mathrm{OPW},\boldsymbol{k}}(\boldsymbol{r}) = \mathrm{e}^{i\boldsymbol{k}\cdot\boldsymbol{r}} + \sum_m b_{mk} \psi^t_{mk} \qquad (5.3)$$

If $b_{mk} = -\int (\psi^t_{mk})^* e^{i\boldsymbol{k}\cdot\boldsymbol{r}}\, d^3\boldsymbol{r}$, it is easy to show that this function will indeed be orthogonal to all the core states ψ^t_{mk}: the contribution to $\int \phi(\psi^t_{mk})^*\, d^3\boldsymbol{r}$ from the second term in (5.3) just cancels the contribution from the first.

The OPW (5.3) has much the same character as an APW – atomic-like within each ion core, where the ψ^t_{mk} terms are large, and plane-wave-like outside, where they are small. Just as with APWs, we can use a set of OPWs to form a Bloch wave

$$\psi_{nk}(\boldsymbol{r}) = \sum c_{\boldsymbol{K}} \phi_{\mathrm{OPW},\boldsymbol{k}-\boldsymbol{K}}(\boldsymbol{r}) \tag{5.4}$$

and use a variational method to choose the best coefficients $c_{\boldsymbol{K}}$; just as with APWs, the resulting set of equations has the form (4.24), and is readily solved on a computer. The advantage of the OPW method is that it is not necessary to use the muffin-tin approximation for $V(\boldsymbol{r})$; the disadvantage is that it works well only if the wave-functions in the solid fall into two clearly distinct groups: the core wave-functions, well approximated by (5.1), and the conduction-electron (or valence-electron) wave-functions, much less tightly bound and well represented by a small number of OPWs.

Although these three methods (and others) were developed in the years 1937–54, few realistic band-structure calculations were attempted until about 1960, when the advent of large computers made them less prohibitively laborious. A few years earlier, in 1957, it began to be possible to determine FS shapes experimentally, and in 1958 it was found by A. V. Gold that the FS of Pb was remarkably similar to that predicted by the NFE model. This was very surprising: it had always been supposed that the NFE model – assuming all the coefficients $V_{\boldsymbol{G}}$ in (4.24) to be small – had little or no relevance to real metals.

It was soon realized that except for the transition metals and the rare-earth metals, which have unfilled d or f shells, the NFE model in fact works remarkably well: the band structure and the shape of the FS can be quite closely reproduced by choosing the right values for a few coefficients $V_{\boldsymbol{G}}$ in (4.24) and setting the rest equal to zero. These $V_{\boldsymbol{G}}$ are in fact the Fourier coefficients of a rather weak and smoothly-varying 'pseudopotential' $V_{\mathrm{PS}}(\boldsymbol{r})$ for which the conduction electron band structure $\varepsilon_n(\boldsymbol{k})$ happens to be much the same as for the (far stronger) potential $V(\boldsymbol{r})$. The corresponding wave-functions

$$\psi_{\mathrm{PS}}(\boldsymbol{r}) = \sum c_{\boldsymbol{K}} e^{i(\boldsymbol{k}-\boldsymbol{K})\cdot\boldsymbol{r}} \tag{5.5}$$

vary smoothly through the unit cell, and show none of the detailed structure that the true wave-functions show inside the ion cores. In fact, it can be shown that if $V_{PS}(\boldsymbol{r})$ is chosen appropriately, the coefficients $c_{\boldsymbol{K}}$ in (5.5) are the same as those in (5.4), and $\psi_{PS}(\boldsymbol{r})$ is effectively an 'unorthogonalized' version of the OPW solution (5.4). What is completely missing from this 'pseudometal' is the set of tightly-bound core functions ψ^t: the whole complex structure of nucleus + core electrons has been replaced by a pseudo-ion with a relatively weak pseudopotential, surrounded only by the outermost valence electrons.

The pseudopotential approach, like the OPW approach, therefore fails for the transition metals and the rare-earth metals, because in these metals there is no clear dividing line between tightly-bound core electrons and loosely-bound valence electrons. Correspondingly, the NFE approximation is no use as a guide to the band structure of these metals, though it is very useful for the remaining metals – the 'simple' metals, as they are called.

What does the band structure look like, for an fcc crystal say, in the NFE approximation? As Fig. 5.2 shows, it looks pretty

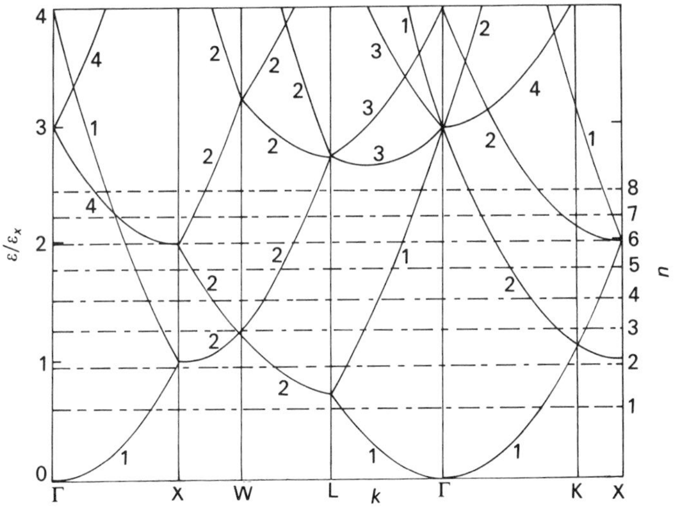

Fig. 5.2 The energy band structure of a free electron fcc metal, plotted along the lines joining the points Γ, X etc. shown in Fig. 5.3. The number against each curve shows its degeneracy. [After Herman (1967) in *An Atomistic Approach to the Nature and Properties of Materials* (ed. Pask), Wiley, New York.]

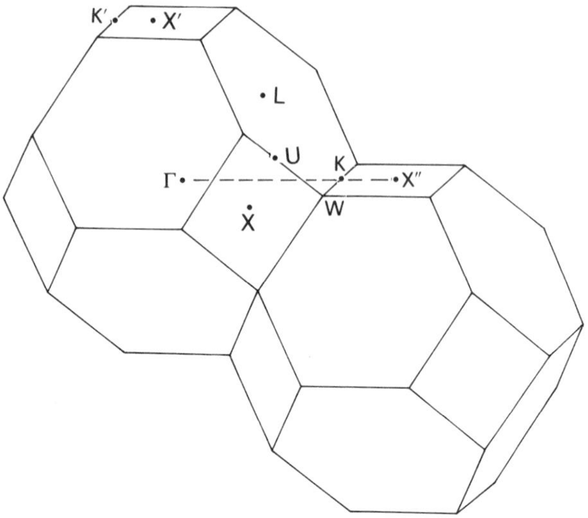

Fig. 5.3 The fcc BZ, showing the standard labelling of symmetry points, and showing that the line KX″ is an extension of the line ΓK.

complicated, at first sight. This is partly because Fig. 5.2 consists of six different $\varepsilon_n(\boldsymbol{k})$ graphs side by side, along six different lines in the BZ, between the points shown in Fig. 5.3. Each curve in Fig. 5.2 is a parabola of the form $\varepsilon = (\hbar^2/2m)(\boldsymbol{k} - \boldsymbol{K})^2$, centred on a different reciprocal lattice (RL) point $\boldsymbol{K}$. For the lowest parabolas, starting out from the point Γ at the centre of the BZ, $\boldsymbol{K} = 0$. If the coordinates of the point X are (1,0,0) (in suitable units), the lowest parabola along ΓX has the form $\varepsilon = \varepsilon_x k_x^2$, rising from 0 at $k_x = 0$ (point Γ) to ε_x at $k_x = 1$ (point X). The next RL point in the ΓX direction is at (2,0,0), and this gives rise to the next higher parabola, $\varepsilon = \varepsilon_x(2 - k_x)^2$. At the point $(\frac{1}{2},0,0)$ this crosses a set of four coincident parabolas: any point on the line ΓX is equidistant from the four RL points (111), $(1\bar{1}1)$, $(11\bar{1})$, $(1\bar{1}\bar{1})$ (where, as usual, $\bar{1}$ denotes -1), and each of these four points gives rise to a parabola of the form $\varepsilon = \varepsilon_x[(1 - k_x)^2 + 1^2 + 1^2]$. This fourfold degeneracy exists only along the symmetry line ΓX; if we move a little way off this line, the distances to the four RL points will become different, and the four parabolas will no longer coincide.

It is not too difficult to trace out in this way the origin of the various curves in Fig. 5.2. Note that the curves from Γ to K continue on smoothly to X, which may seem odd, because in Fig. 5.3 X is 'round the corner' from K. But the ΓK axis in fact passes through the point X″, at the centre of one of the square faces of the adjacent BZ (remember that, by definition, the BZ is bounded on all sides by other BZs), and what in fact happens is that the $\varepsilon_n(\boldsymbol{k})$ curves continue smoothly from Γ through K to X″. The line KX″ in the next BZ is equivalent to the line K′X′ in the central BZ, and this in turn is equivalent, because of cubic symmetry, to the line UX. So the points U and K are equivalent, and the last part of Fig. 5.2 should be thought of as showing $\varepsilon_n(\boldsymbol{k})$ along KX″ or along UX, rather than along KX.

The scale on the right of Fig. 5.2 shows the number of conduction electrons n that can be accommodated, per real-space unit cell, for various values of $\varepsilon_F/\varepsilon_x$, where ε_F is the Fermi energy. In the NFE approximation,

$$k_F/k_0 = (3n/2\pi)^{1/3} \tag{5.6}$$

and hence

$$\varepsilon_F/\varepsilon_x = (3n/2\pi)^{2/3} \tag{5.7}$$

where k_F is the radius of the free-electron Fermi sphere and k_0 is the distance ΓX (problem 5.1). In the NFE approximation, we can thus visualize the FS by thinking of a sphere of radius k_F centred on each reciprocal lattice point, with the spheres reconnected wherever they intersect, as described in section 4.3.

For small n, the spheres do not intersect at all, and the FS is therefore spherical. At $n \approx 1.4$, Fig. 5.2 shows that the FS touches the BZ boundary at the points L, so that for $n > 1.4$ a section through the FS in the (111) direction will look like Fig. 4.3; in the lower band, the spheres join together, and the segments overlapping into the second band form lens-shaped surfaces, centred on the points L. Since contact occurs simultaneously at all eight points L, the FS in the first band now consists (in the periodic zone scheme) of a network of spheres each of which is linked to all its eight neighbours across the (111) faces of the BZ.

For $n \approx 2$, Fig. 5.2 shows that contact occurs at the points X, so that for $n > 2$ the spheres are linked to their remaining neighbours across the X faces too, and lenses appear in the second band around X. Then for $n \approx 2.5$, contact occurs at K and U, so that the second-band lenses link up into a single surface, and all that is left of the spheres in the first band is a set of isolated patches around the points W, which join up in the periodic zone scheme to form small 'hole' pockets – pockets of

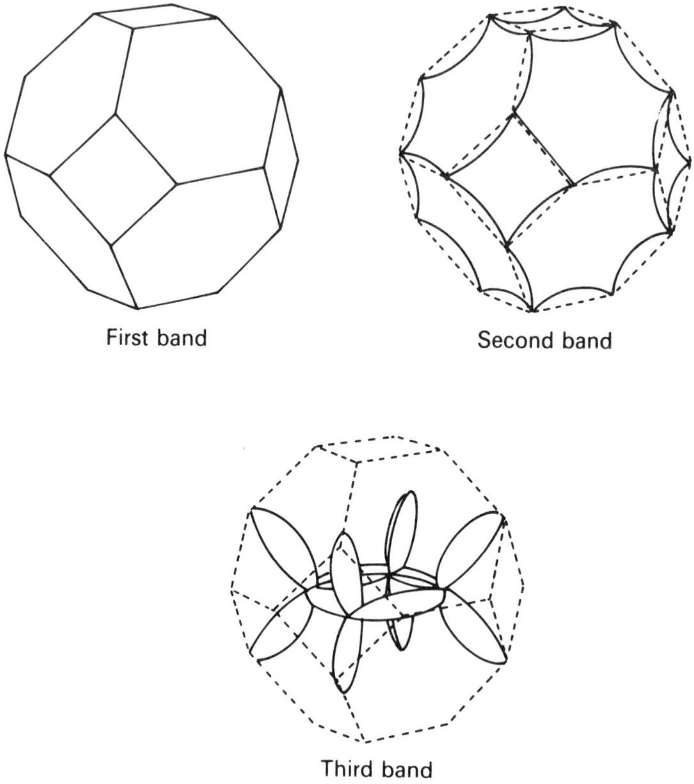

Fig. 5.4 The FS for a free-electron metal with three electrons per unit cell. In the third band, the centre of the figure is at the point X′ of Fig. 5.3, not the point Γ. [From Harrison (1960), *Phys. Rev.* **118**, 1190.]

empty states, surrounded by filled states. Finally, at n just less than 3, the spheres pass out of the BZ altogether, at points W, and these small hole pockets disappear, leaving the first band completely full.

The resulting FS for $n = 3$ is shown in Fig. 5.4. Very much as in Fig. 4.5, the first band is full, and the second contains a large 'hole' surface, with filled states outside and empty states inside, formed by the joining-up of the lenses which existed (for smaller n) around L and X. Again as in Fig. 4.5, the filled states overlap into bands 3 and 4: in band 3 the FS consists of a set of thin tubes running along the edges of the BZ to form a three-dimensional network, and in band 4 (not shown) there are tiny pockets of filled states around the corners W.

Now let us compare these NFE predictions with reality. Al is a

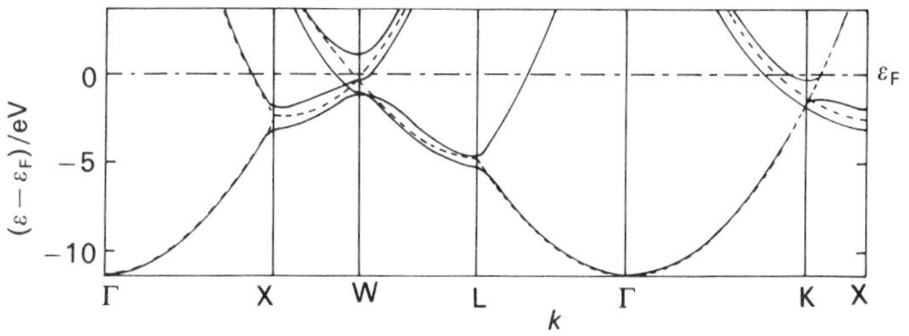

Fig. 5.5 The band structure of Al (full lines) compared with the NFE approximation of Fig. 5.2 (broken lines). [After Segall (1961), *Phys. Rev.* **124**, 1797.]

simple fcc metal, with three valence electrons per atom, so that $n = 3$. Figure 5.5 shows that its band structure, from an accurate KKR calculation, is indeed remarkably close to the predictions of the NFE model, shown dashed. Correspondingly, the FS of Al, as determined experimentally, is very close to that shown in Fig. 5.4. The slight differences that do exist can be accurately modelled by a weak pseudopotential with only two sets of components V_G:0.24 eV for the $\boldsymbol{G} = \{111\}$ set of lattice vectors, and 0.76 eV for the $\{200\}$ set. The cusp-like edges of the second-band surface are rounded off slightly, as we should expect, and the tiny pockets of electrons in band 4 are absent, because the energy here is very slightly above ε_F, instead of being very slightly below. The network of tubes along the edges of the BZ in band 3 is broken up into a set of narrow square rings around the edges of the square faces, which do not quite join up at the points W. Apart from these minor changes, the NFE model predicts the FS of Al remarkably closely; and it does almost as well for many of the other simple metals.

The noble metals Cu, Ag and Au lie between the simple metals and the transition metals in the periodic table, and their band structure, as shown in Fig. 5.6, is likewise intermediate. Some of the branches clearly derive from the NFE bands of Fig. 5.2, shown dashed, but there are also five almost horizontal bands running a little way below ε_F, not accounted for at all on the NFE model. These bands arise from the ten d-shell electrons in the free atom. In simple metals, these bands would be so tightly bound that they would lie below the bottom of the diagram; in transition metals they would be higher in energy, straddling ε_F, so that they would be only partly filled. The FS would then include contributions from these partly-filled d-bands. In the

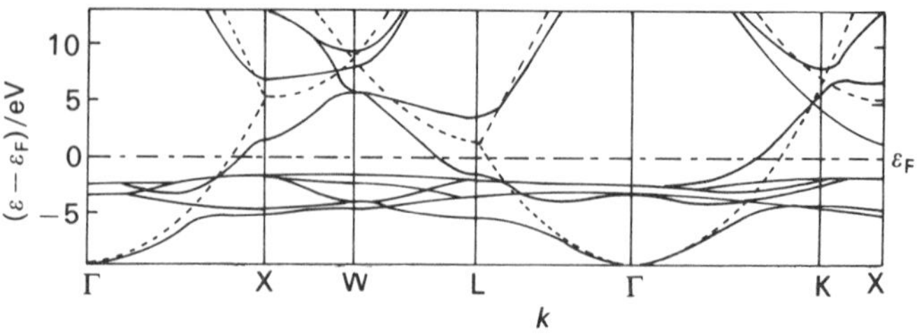

Fig. 5.6 The band structure of Cu (full lines) compared with the NFE approximation (broken lines). [After Jepsen *et al.* (1981), *Phys. Rev.* **B23**, 2684.]

noble metals, the d-bands do not lie high enough to contribute to the FS, which is therefore not too different from the NFE prediction.

Since these metals are monovalent, the NFE FS would be simply a sphere, coming close to the BZ boundary at point L, but not quite touching it. But Fig. 5.6 shows that in Cu (and in Ag and Au too) none of the $\varepsilon_n(\boldsymbol{k})$ curves along ΓL passes through the energy ε_F. The three NFE parabolas rising from Γ towards X, L and K encounter and mix with the d-bands on the way, but whereas the other two emerge from the encounter in time to pass through ε_F before reaching the BZ boundary, the ΓL parabola does not – the energy at L stays below ε_F.

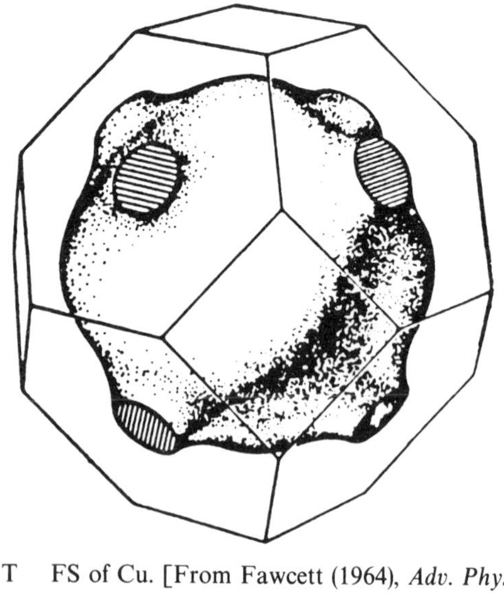

Fig. 5.7 T FS of Cu. [From Fawcett (1964), *Adv. Phys.* **13**, 139.]

The point L therefore lies below the FS, because the states at L have $\varepsilon < \varepsilon_F$, and indeed Fig. 5.6 shows that we have to go some little way from L along LW before reaching the point where the $\varepsilon(\boldsymbol{k})$ curve rises through ε_F; at this point, the FS contacts the BZ boundary.

The resultant form of the FS is shown in Fig. 5.7. In the periodic zone scheme, it becomes a set of spheres connected by necks across the (111) faces of the BZ to become an unbounded or 'open' surface, extending throughout $\boldsymbol{k}$-space. The experimental determination of this FS, by Pippard in 1957, marked the beginning of our detailed knowledge of the band structures of metals.

5.2 METALS, SEMICONDUCTORS AND INSULATORS

So far in this chapter we have talked almost entirely about metals. Semiconductors and insulators differ from metals, from a band-structure viewpoint, in just two respects: there is an energy gap in the band structure, with some bands lying below the gap and others above it; and the Fermi energy ε_F lies in this gap. At $T = 0$, all the bands below the gap are therefore completely full, and all the bands above the gap are completely empty. For this to be possible, the material must contain an even number of electrons per unit cell, as (4.9) shows, and the energy splitting between bands must be big enough for the highest energy state in the lower band to lie below the lowest energy state in the higher band. There is no qualitative difference between semiconductors and insulators; in semiconductors, the energy gap is simply narrower than it is in insulators.

Perhaps surprisingly, the band structure of a semiconductor like Si, or even of an insulator like diamond, is not too different from what the NFE model would predict, as Fig. 5.8 shows for Si. The highest energy state in the valence band (as the band below the gap is called) is at Γ, and the lowest energy state in the conduction band (above the gap) is at a point near X on the ΓX axis; by symmetry, there are six such points. The energy gap between these two states is about 1.1 eV. Since the Si atom has four valence electrons, and the crystal contains two atoms per unit cell, there are eight valence electrons per unit cell, and at $T = 0$ these just fill the four valence bands below the gap. (Two of these four bands are degenerate along ΓL and ΓX.) For $T > 0$, some electrons are thermally excited across the gap into the states near X, leaving vacant 'hole' states around Γ in the valence band. The constant-energy surfaces around the conduction-band minima, near X, are cigar-shaped, elongated along ΓX. By symmetry there are six

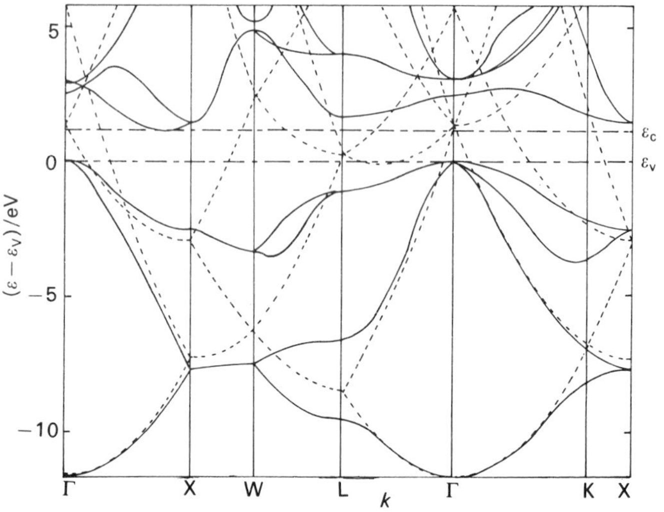

Fig. 5.8 The band structure of Si (full lines) compared with the NFE approximation (broken lines). [After Stohr and Bross (1978), *Phys. Stat. Solidi*, **B90**, 497.]

such minima, and six sets of cigar-shaped surfaces around them. The band structure near Γ in the valence band is in fact a bit complicated, because there are actually three (almost) degenerate bands there, and we return to this in section 13.1.

In Ge, another semiconductor with the same crystal structure as Si, the lowest energy states in the conduction band are at L, in the middle of the hexagonal BZ faces, and the constant-energy surfaces around the points L are again cigar-shaped, pointing towards Γ. Although there are eight points L on the surface of the BZ, opposite points are of course equivalent, so that there are in fact only four distinct energy minima, and four sets of cigar-shaped surfaces around them.

In other semiconductors such as InSb and GaAs, there is only one minimum-energy point in the conduction band, at Γ, so that the surrounding energy surfaces are to good approximation spherical, like the surfaces around the valence-band maximum, which is also at Γ.

5.3 DENSITY OF STATES AND HEAT CAPACITY

For free electrons, the density of states is given by (1.26), and we have $g(\varepsilon) \propto \varepsilon^{1/2}$. In the NFE approximation, $g(\varepsilon)$ is still given by (1.26), because the total volume of $\boldsymbol{k}$-space between energy ε and $\varepsilon + \delta\varepsilon$ remains unchanged; it has merely been redistributed between the different bands (compare Figs 4.5 and 4.6), so that $g(\varepsilon) = \sum g_n(\varepsilon)$, summed over all bands n. Within any one band, however, the total number of states must be given by

$$\int g_n(\varepsilon)\,\mathrm{d}\varepsilon = 2N \tag{5.8}$$

where the integral runs from $\varepsilon_{\min}$, the lowest energy in the band, to $\varepsilon_{\max}$, the highest, and N is the number of unit cells per unit volume in the crystal (cf. (4.9)). Moreover, the result (5.8) must be true generally, and not just in the NFE approximation. Between $\varepsilon_{\min}$ and $\varepsilon_{\max}$, $g(\varepsilon)$ must rise from zero, go through a peak, and fall again to zero. The way it does so is shown, schematically, in Fig. 5.9. The sudden breaks in slope, called van Hove singularities, occur where the constant-energy surfaces change their topology. In discussing Fig. 5.2, we saw that as the number of electrons n was increased, the shape of the FS in the lowest band changed from a sphere to a linked set of spheres, and then to a set of hole surfaces around the zone corners. Not surprisingly, the van Hove singularities occur at the energies where these sudden changes in the geometry of the energy surfaces occur. In Fig. 5.9, we have shown the simplest possible case, where there are only two such sudden changes, apart from those at the end-points $\varepsilon_{\min}$ and $\varepsilon_{\max}$. There cannot be less than two, and often there are more (problem 5.2).

In a metal, different bands overlap in energy, and the densities $g_n(\varepsilon)$ likewise overlap, as shown in Fig. 5.9a. In a semiconductor or insulator, the energy gap shows up as a gap in the density of states, as in Fig. 5.9b. It follows from (2.3) that the electronic heat capacity C_{el}

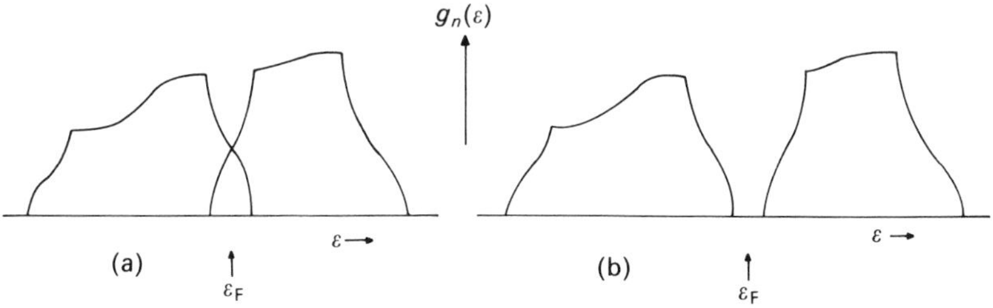

Fig. 5.9 Showing schematically the van Hove singularities in $g_n(\varepsilon)$, and showing how the bands overlap in metals but not in semiconductors.

will be zero in an insulator, and negligible in a semiconductor, since ε_F lies in the gap, where $g(\varepsilon)$ is zero.

In a simple metal, for which the NFE model is a reasonable approximation, C_{el} should be not too different from the free-electron value. In a transition metal, however, the presence of the d-bands may make C_{el} much larger. We can see why from Fig. 5.9. The d-bands are relatively tightly bound, and therefore narrow; that is, $\varepsilon_{max} - \varepsilon_{min}$ is small. Whenever a band is narrow, $g_n(\varepsilon)$ must be large within the band, if (5.8) is to be satisfied. In Cu, the d-bands lie below the FS, and do not contribute to C_{el}, but in the transition metals, the FS usually lies in the middle of the d-bands, and C_{el} is correspondingly large. There are also other and more subtle effects which can increase C_{el}, and we shall come back to these in Chapter 6, after discussing the form of $V(\mathbf{r})$.

6

The potential $V(r)$; many-body effects

6.1 THE ONE-ELECTRON APPROXIMATION AND THE CHOICE OF $V(\boldsymbol{r})$

In discussing the solution of the Schrödinger equation (4.1), we have so far simply treated $V(\boldsymbol{r})$ as a known periodic function of $\boldsymbol{r}$. But in fact, the choice of $V(\boldsymbol{r})$ is itself a subtle and difficult problem, because (4.1) is only an approximation to a much more complicated equation, and $V(\boldsymbol{r})$ has to be chosen in such a way as to make the approximation as good as possible. It is worth looking briefly at this more complicated equation, just to see what approximations are needed in order to reduce it to the soluble form (4.1).

We start by regarding the solid as an assembly of n nuclei (each having charge Ze) and nZ electrons, and writing the Schrödinger equation for this assembly in the form

$$(K_{\mathrm{e}} + K_{\mathrm{p}} + V_{\mathrm{ep}} + V_{\mathrm{pp}} + V_{\mathrm{ee}})\Psi = \varepsilon_{\mathrm{T}}\Psi \tag{6.1}$$

where ε_{T} is the total energy of the whole assembly, and $\Psi = \Psi(\boldsymbol{r}_1, \ldots \boldsymbol{r}_i, \ldots \boldsymbol{r}_{nZ};\ \boldsymbol{R}_1, \ldots \boldsymbol{R}_{\mathrm{I}}, \ldots \boldsymbol{R}_n)$ is a function (an exceedingly complicated function) of the coordinates $\boldsymbol{r}_i$ ($i = 1$ to nZ) of all the nZ electrons and $\boldsymbol{R}_I$($I = 1$ to n) of all the n nuclei. (In fact Ψ will also depend on the spin 'coordinate' s_i of each electron, i.e. whether the spin is up ($s_i = \uparrow$) or down ($s_i = \downarrow$), but for simplicity we shall only include this complication where we have to, as in equations (6.6)–(6.9).) The total kinetic energy of the electrons, K_{e}, and of the positive nuclei, K_{p}, is given by the operators

$$K_{\mathrm{e}} = -(\hbar^2/2m)\sum_i \nabla_i^2; \quad K_{\mathrm{p}} = -(\hbar^2/2M)\sum_i \nabla_I^2 \tag{6.2}$$

where $\nabla_i^2 = \partial^2/\partial x_i^2 + \partial^2/\partial y_i^2 + \partial^2/\partial z_i^2$, etc. and M is the mass of a nucleus. The potential energy terms V_{ep}, etc. are given by

$$V_{\text{ep}} = -\sum_i \sum_J \frac{Ze^2}{|\boldsymbol{r}_i - \boldsymbol{R}_J|}; \quad V_{\text{pp}} = \frac{1}{2} \sum_{I \neq J} \sum_J \frac{Z^2 e^2}{|\boldsymbol{R}_I - \boldsymbol{R}_J|};$$

$$V_{\text{ee}} = \frac{1}{2} \sum_{i \neq j} \sum_j \frac{e^2}{|\boldsymbol{r}_i - \boldsymbol{r}_j|} \tag{6.3}$$

where we have for brevity omitted the factors $4\pi\varepsilon_0$, and where the three terms represent respectively the Coulomb attraction between the electrons and the nuclei, the mutual repulsion of the nuclei, and the mutual repulsion of the electrons. (As usual, the $\frac{1}{2}$ in V_{pp} and V_{ee} is needed to avoid counting each interaction twice over.) The first of these terms can be written as

$$V_{\text{ep}} = \sum_i V_{i\text{p}}, \quad \text{where} \quad V_{i\text{p}} = -\sum_J \frac{Ze^2}{|\boldsymbol{r}_i - \boldsymbol{R}_J|} \tag{6.4}$$

Because the nuclei are far heavier than the electrons, they move far more slowly, and we can for many purposes neglect their motion and think of the electrons as moving through an effectively stationary assembly of nuclei. In this ('adiabatic') approximation, we can neglect the terms K_{p} and V_{pp} in (6.1), which then reduces to an equation for the many-electron wave-function $\Psi(\boldsymbol{r}_1, \ldots \boldsymbol{r}_{nZ})$:

$$\left(K_{\text{e}} + \sum_i V_{i\text{p}} + V_{\text{ee}} \right) \Psi = \varepsilon_{\text{e}} \Psi \tag{6.5}$$

If it were not for the electron–electron interaction term V_{ee}, (6.5) could be split up into a set of one-electron wave equations like (4.1), each satisfied by a one-electron wave-function $\psi(\boldsymbol{r})$, and the problem would be soluble, at least in principle. It is this interaction term, representing the Coulomb repulsion between the electrons, which makes the problem insoluble without drastic approximations. Qualitatively, though, we can see easily enough what the effect of V_{ee} will be. The Coulomb repulsion will produce 'correlation effects': the electrons will tend to avoid one another (that is, $|\Psi|^2$ will become small if the separation $|\boldsymbol{r}_i - \boldsymbol{r}_j|$ between any pair of electrons becomes small), so that each electron is surrounded by a 'correlation hole' which moves around with it. It is this correlation hole which is lost if we try to approximate Ψ by a combination of one-electron wave-functions $\psi(\boldsymbol{r})$.

There is one important conclusion we can draw about the form of the many-electron wave-function, before we start approximating. Because the electrons are indistinguishable, the probability $|\Psi|^2$ of finding the electron assembly with particular values of $\boldsymbol{r}_1, \cdots \boldsymbol{r}_i \cdots \boldsymbol{r}_j \cdots \boldsymbol{r}_{nZ}$ must remain the same if the positions of any two electrons, *i* and *j* say, are interchanged; otherwise it would mean that the probability of finding the system with an electron at $\boldsymbol{r}_i$, say, and another at $\boldsymbol{r}_j$, depended on *which* electron was at $\boldsymbol{r}_i$ and which was at $\boldsymbol{r}_j$, and this would not make sense physically. If $|\Psi|^2$ is to remain the same, Ψ itself must either remain the same or at most change sign. It turns out that for bosons – particles of integral spin – Ψ stays the same, while for fermions, including electrons, Ψ changes sign when two particles are interchanged.

If we interchange two spin-up electrons, this means that

$$\Psi(\ldots \boldsymbol{r}_A\uparrow, \ldots \boldsymbol{r}_B\uparrow, \ldots) = -\Psi(\ldots \boldsymbol{r}_B\uparrow, \ldots \boldsymbol{r}_A\uparrow, \ldots), \qquad (6.6)$$

while if we interchange a spin-up and a spin-down electron,

$$\Psi(\ldots \boldsymbol{r}_A\uparrow, \ldots \boldsymbol{r}_B\downarrow, \ldots) = -\Psi(\ldots \boldsymbol{r}_B\downarrow, \ldots \boldsymbol{r}_A\uparrow, \ldots). \qquad (6.7)$$

Consider now what happens when $\boldsymbol{r}_A = \boldsymbol{r}_B$, so that interchanging the two electrons reduces to interchanging their spins. In (6.7), this presents no problems; there is no reason why Ψ should not change sign when an up spin and a down spin are interchanged. But in (6.6), the interchange leaves the whole set of variables completely unaffected, if $\boldsymbol{r}_A = \boldsymbol{r}_B$. Just as a simple function $f(x)$ can only satisfy $f(x) = -f(x)$ if $f(x) = 0$, so here the more complicated function Ψ can only satisfy $\Psi(\ldots \boldsymbol{r}_A\uparrow, \ldots \boldsymbol{r}_A\uparrow, \ldots) = -\Psi(\ldots \boldsymbol{r}_A\uparrow, \ldots \boldsymbol{r}_A\uparrow, \ldots)$ if $\Psi = 0$. It follows, then, that the probability $|\Psi|^2$ of finding two parallel-spin electrons at the same point $\boldsymbol{r}$ must vanish. And clearly $|\Psi|^2$ cannot vanish discontinuously at $\boldsymbol{r}_A = \boldsymbol{r}_B$; it must fall smoothly to zero as $\boldsymbol{r}_A$ approaches $\boldsymbol{r}_B$. It follows that every electron in the system is surrounded by a region – the 'Pauli hole' or 'exchange hole' – in which we are unlikely to find another electron of the same spin.

Note that this effect is nothing to do with the Coulomb repulsion between the electrons, which produces the correlation hole. The Coulomb repulsion reinforces the exchange effect in keeping parallel-spin electrons apart, and is the only effect keeping antiparallel ones apart; but the exchange effect would still keep parallel spins apart even if the particles were uncharged: it arises simply from the antisymmetry condition (6.6).

This antisymmetry condition is the basic form of the Pauli exclusion principle. If we try to approximate the many-electron wave-function Ψ by a combination of one-electron wave-functions $\psi_i(\boldsymbol{r}_i, s_i)$, the simplest possible combination would be

$$\Psi = \psi_1(\boldsymbol{r}_1, s_1)\psi_2(\boldsymbol{r}_2, s_2)\psi_3(\boldsymbol{r}_3, s_3)\ldots \tag{6.8}$$

but this does not satisfy the antisymmetry condition. The simplest combination that does satisfy this condition is the *Slater determinant*

$$\Psi = \begin{vmatrix} \psi_1(\boldsymbol{r}_1, s_1) & \psi_1(\boldsymbol{r}_2, s_2) & \psi_1(\boldsymbol{r}_3, s_3) & \cdots \\ \psi_2(\boldsymbol{r}_1, s_1) & \psi_2(\boldsymbol{r}_2, s_2) & \psi_2(\boldsymbol{r}_3, s_3) & \cdots \\ \psi_3(\boldsymbol{r}_1, s_1) & \psi_3(\boldsymbol{r}_2, s_2) & \psi_3(\boldsymbol{r}_3, s_3) & \cdots \\ \vdots & \vdots & \vdots & \end{vmatrix} \tag{6.9}$$

and this at once leads to the more familiar form of the exclusion principle, that no two electrons can occupy the same state with the same spin (problem 6.1).

So far we have thought of the many-electron wave-function Ψ of (6.5) as describing *all* the nZ electrons in the system, moving in the field of the n nuclei and of each other. The first obvious approximation we can make in trying to solve (6.5) is to think of the inner electrons around each nucleus as tightly bound to it to form an ion core which is unaffected by the neighbouring ion cores, and to include in Ψ only the outer valence electrons, which are not bound to one or other of the ion cores. The same equation (6.5) can still be used to describe the problem, but V_{ip} now represents the potential energy of electron i in the field of the ion cores rather than of the nuclei alone. Even this approximation is not a simple one, though, because there will still be exchange and correlation effects between the valence electrons and those in the ion cores, and V_{ip} has to be chosen carefully to take account of these.

The crudest approximation we can make in trying to solve (6.5) is to write Ψ as a simple product of one-electron wave-functions of the form (6.8). Each of the wave-functions $\psi(\boldsymbol{r})$ then has to satisfy an equation of the form (4.1):

$$-(\hbar^2/2m)\nabla^2\psi + V_{\mathrm{p}}(\boldsymbol{r})\psi + V_{\mathrm{e}}(\boldsymbol{r})\psi = \varepsilon\psi \tag{6.10}$$

where we have written V_{ip} as $V_\mathrm{p}(\mathbf{r})$, and $V_\mathrm{e}(\mathbf{r})$ represents the potential energy of the electron in the *averaged-out* field of all the other electrons:

$$V_\mathrm{e}(\mathbf{r}) = e^2 \int \mathrm{d}^3\mathbf{r}'\, n(\mathbf{r}')/|\mathbf{r}-\mathbf{r}'| \tag{6.11}$$

where

$$n(\mathbf{r}') = \sum |\psi(\mathbf{r}')|^2 \tag{6.12}$$

summed over all other electrons. In practice we can sum over *all* electrons, because the difference of one among 10^{23} is negligible, and $n(\mathbf{r})$ is then simply the electron density at point $\mathbf{r}$. (If V_p is the potential due to the ion cores, rather than that due to the bare nuclei, V_e will be the potential due to the remaining electrons – those outside the ion cores – and correspondingly the sum in (6.12) extends only over these electrons.) To find $\psi(\mathbf{r})$ for *one* electron by solving (6.10), we thus need first to know $\psi(\mathbf{r})$ for *all* electrons, so that we can find $n(\mathbf{r}')$ and hence $V_\mathrm{e}(\mathbf{r})$. This is not as bad as it sounds – by starting with a set of trial functions $\psi(\mathbf{r})$ and iterating a few times, we arrive quite rapidly at a set of self-consistent solutions.

This *Hartree approximation*, crude though it is, in fact works surprisingly well. We might hope to do better by starting from the antisymmetric wave-function (6.9) instead of (6.8), but in fact this *Hartree–Fock approximation*, disappointingly, works rather less well than Hartree, at least in metals. It is also very much more difficult to solve, because a complicated 'exchange term' now has to be added to V_e in (6.10).

Neither the Hartree nor the Hartree–Fock approximation includes the detailed correlation effects produced by the V_ee term in (6.5), and this is very difficult to do. Of many alternative approaches to the problem, the simplest and most successful is the 'density functional' method of Hohenberg, Kohn and Sham, so called because the density of electrons $n(\mathbf{r})$ plays a central role.

What determines $n(\mathbf{r})$? If, for a given form of periodic ionic potential $V(\mathbf{r})$, we could solve (6.5) exactly for the many-electron wave-function Ψ, we could then find $n(\mathbf{r})$ from

$$n(\mathbf{r}) = N \int |\Psi|^2\, \mathrm{d}^3\mathbf{r}_2\, \mathrm{d}^3\mathbf{r}_3 \mathrm{d}^3\mathbf{r}_4 \ldots \mathrm{d}^3\mathbf{r}_N \tag{6.13}$$

integrated over all but one of the N sets of coordinates of the N

electrons. What is less obvious is the fact that a given form of $n(\boldsymbol{r})$ is compatible with only one form of $V_{\mathrm{p}}(\boldsymbol{r})$, and hence with one solution $\Psi(\boldsymbol{r}_1,\boldsymbol{r}_2,\ldots)$. In other words, V_{p} and Ψ can be thought of as 'functionals' of $n(\boldsymbol{r})$. (A functional is simply a function of something which is itself a function.)

Now imagine our assembly of interacting electrons replaced by an equivalent assembly of *non-interacting* electrons, so that $V_{\mathrm{ee}}=0$ in (6.5), and at the same time let the potential V_{p} be replaced by an effective potential V_{eff}, whose form is yet to be determined. Then, just as in the Hartree approximation, (6.5) can be factorized into a set of one-electron equations very much like (6.10):

$$-(\hbar^2/2m)\nabla^2\psi + V_{\mathrm{eff}}(\boldsymbol{r})\psi = \varepsilon\psi \tag{6.14}$$

From the solutions $\psi(\boldsymbol{r})$ to this one-electron equation we can at once find the electron density from (6.12). Finally, we can then choose the form of $V_{\mathrm{eff}}(\boldsymbol{r})$ so that this electron density is equal to the exact density (6.13), given by the exact many-electron wave-function. The resultant set of one-electron wave-functions will then be, in a well-defined sense, the best possible approximation to the exact solution and, in particular, the total energy of the system should be given correctly by this set of solutions.

Since we do not in fact know the exact density, we cannot actually carry out this programme; nevertheless, this argument does provide us with a clear logical route from the insoluble many-electron equation (6.5) to the far simpler one-electron equation (6.14). With reasonable assumptions about V_{eff} and its relationship to $n(\boldsymbol{r})$, it yields remarkably accurate results for the one-electron energies, when compared with experiment. This density functional approach now forms the basis of most band-structure calculations. If the ionic lattice is periodic, V_{eff} will also be periodic, so that all we said about (4.1) and its solutions applies equally well to (6.14); in particular, the wave-functions are Bloch waves, $\psi_{n\boldsymbol{k}}(\boldsymbol{r})$, with energies $\varepsilon_n(\boldsymbol{k})$.

It turns out that the energies $\varepsilon_n(\boldsymbol{k})$ do not differ greatly from those given by the Hartree approximation (6.10). This is because, if we write V_{eff} in (6.14) in the form

$$V_{\mathrm{eff}} = V_{\mathrm{p}} + V_{\mathrm{e}} + V_{\mathrm{xc}} \tag{6.15}$$

the term V_{xc}, which represents the effects of exchange and correlation, turns out to be rather small. One might have expected the Coulomb repulsion between electrons to have a much larger effect, because the potential energy of repulsion between two electrons at a separation of 1 or 2 Å (0.1 or 0.2 nm) is about 10 eV. But in a metal, this interaction is much reduced by *screening* effects: because the metal as a whole is electrically neutral, the lower density of other electrons in the 'correlation hole' means that each electron is effectively surrounded by a positively charged screening cloud, and the net effect is to produce a potential which falls off exponentially, rather than simply as $1/r$. (As we shall see in section 8.1, charged *impurities* are screened in much the same way.)

6.2 MANY-BODY EFFECTS

For many purposes, a one-electron approximation of the form (6.14) works very well. When it fails to do so, we speak of 'many-body' effects. For example, by appropriate choice of V_{eff} we can take good account of electron–electron interactions, provided that the electron assembly is in its ground state, with all states $\varepsilon(\boldsymbol{k})$ filled up to the Fermi energy ε_{F}, and all higher states empty. But at finite temperatures, or when an electric field is applied, some of the states below ε_{F} will be empty, and some of those above ε_{F} will be occupied. The system is no longer in its ground state, and because of this there may be a change in $n(\boldsymbol{r})$. If so, it follows that there will also be changes in V_{eff} and so in $\varepsilon(\boldsymbol{k})$. This is the basis of Landau's theory of 'Fermi liquids', which historically pre-dated density functional theory, though from the present viewpoint it is an extension of it. We shall not discuss Landau's theory in detail, because it turns out that electron–electron interactions have rather little effect on the properties of metals, under normal conditions. The predicted electronic heat capacity, for example, differs by only 5% or so from the value predicted using the Hartree approximation. But in fact the *observed* heat capacity is often 50% or more greater than this, because of another many-body effect – *electron–phonon* interactions.

Phonons are the natural modes of vibration of the crystal lattice, and we shall have more to say about them, as a source of electron scattering, in section 8.2. Here we are concerned with a different and more subtle kind of electron–phonon interaction. First, though, it is worth noting that (6.1) enables us to work out the phonon spectrum: the relationship between the frequency ω and the wave-vector $\boldsymbol{q}$ of a sinusoidal wave of ionic displacements in the lattice. Suppose that we solve (6.5), not for a perfect crystal but for a crystal in which the ions are displaced slightly from their lattice sites. Then the energy ε_e of the ground state of the electron assembly will be a function of the ionic positions, $\boldsymbol{R}_1, \ldots, \boldsymbol{R}_n$. If we substitute (6.5) in (6.1), we thus have an equation for the motion of the ions:

$$(K_p + V_{pp} + \varepsilon_e)\Psi = \varepsilon_T \Psi \tag{6.16}$$

where $\Psi = \Psi(\boldsymbol{R}_1, \ldots, \boldsymbol{R}_n)$, and the solutions of this equation will give us the phonon spectrum.

But this spectrum is not our primary concern here. Like (6.5), (6.16) still involves the adiabatic approximation – the assumption that the ions move so slowly compared with the electrons that we can treat them as being at rest while discussing the electron motion. This is indeed a very good approximation, because the ions are so much heavier, but they are not infinitely heavy, and as an electron moves through the crystal it tends to *distort* the lattice very slightly, by attracting towards it the ions in its neighbourhood, and this in turn will modify the energy $\varepsilon(\boldsymbol{k})$ of the electron. This lattice distortion can be described in terms of phonons, and the effect is thus due to electron–phonon interactions.

It is beyond the scope of this book to consider these interactions in detail, but it turns out that their effect is only appreciable for electrons near the Fermi surface. If ω_m is the highest frequency in the phonon spectrum of the crystal (so that $\hbar\omega_m \sim k\theta_D$, where θ_D is the Debye temperature), only electrons within about $\pm\hbar\omega_m$ of ε_F will have their energies affected. At ε_F itself, the energy is unaffected (so that the shape of the FS is unaffected); below ε_F it is raised somewhat by the interactions and above ε_F it is lowered. The rate of change of $\varepsilon(\boldsymbol{k})$ with k_n, the component of $\boldsymbol{k}$ normal to the Fermi surface, is therefore reduced. As we shall see in section 7.1, the velocity $\boldsymbol{v}(\boldsymbol{k})$ of an electron is proportional to this rate of change, and the density of states – and hence the heat capacity, from (2.3) – is inversely proportional to it, so that the effect is to increase the electronic heat capacity. Detailed calculation shows that when this effect is taken into account, the predicted heat capacities are usually in good agreement with those observed experimentally. Sometimes, though, the experimental values are much greater, and we return to this below.

There is one further and still more subtle effect produced by these electron–phonon interactions: the phenomenon of superconductivity. This phenomenon – the sudden vanishing of electrical resistivity in some metals as they are cooled down below a transition temperature T_c, typically around 1–10 K – was discovered experimentally by Onnes in 1911, but remained completely unexplained until 1956. It was then shown by Cooper, and in more detail by Bardeen, Cooper and Schrieffer, that the effect resulted from a 'pairing' of electrons of opposite spin and opposite $\boldsymbol{k}$, brought about by the electron–phonon interaction. Very roughly, the lattice distortion produced by one electron, with energy close to ε_F, exerts an attractive interaction on a second electron, which is strongest when the two electrons are in states $\boldsymbol{k}\uparrow$ and $-\boldsymbol{k}\downarrow$. For $T < T_c$, the two electrons can then lower their energy by forming a 'Cooper pair', and this pairing gives rise to the vanishing of resistance and to other phenomena such as the Meissner effect – the expulsion of magnetic fields from within the material – which are characteristic of superconductivity.

Until 1986, the highest known T_c was 23.2 K in Nb_3Ge, discovered in 1973. In September 1986, Bednorz and Müller in Zürich reported the discovery of a new superconducting material, with composition $(La_{2-x}Ba_x)CuO_{4-y}$ (with $x, y \sim 0.2$), having $T_c \sim 30$ K. Workers in the USA, Japan and China very quickly showed that similar compounds containing Y (or other rare earth metals) in place of the La had still higher values of T_c, and by March 1987 values of T_c of around 95 K were being reported for the compound $YBa_2Cu_3O_7$. Technologically, this could be the most important discovery in physics since the invention of the transistor, and it could have a comparable impact on society, though only if it is possible to produce material which can carry a reasonably high current without losing its superconductivity; so far, this has proved very difficult to achieve.

It has yet to be established what mechanism operates to produce the pairing of electrons in these materials, but it seems highly unlikely that the electron–phonon interaction can be responsible. It is, however, beyond the scope of this book to give a detailed account of superconductivity, or of these extraordinary new developments, and we must leave the subject here.

Just as electron–phonon interactions may not always be the cause of superconductivity, so they may not always be the cause of high electronic heat capacities. In particular, there is a small group of *heavy fermion* materials which have remarkably high values of γ, where $C_{el} = \gamma T$, of around 1 J mole^{-1} K^{-2} (whereas most metals have $\gamma \sim 10^{-2}$–10^{-3} J mole^{-1} K^{-2}). As we saw in discussing electron–phonon effects, a high value of γ implies a small slope $d\varepsilon(\boldsymbol{k})/dk_n$ at the Fermi surface. On a simple model in which $\varepsilon(\boldsymbol{k})$ has the free-electron-like form $\varepsilon(\boldsymbol{k}) = \hbar^2 k^2/2m^*$, with an *effective mass* m^*, we have $d\varepsilon/dk_n \propto 1/m^*$, so that a large γ corresponds to a large effective mass m^* – hence the name 'heavy fermions'. In a typical heavy-fermion system such as UPt_3, m^* turns out to be about $200m$, where m is the free electron mass, far larger than could be accounted for by electron–phonon interactions.

The heavy-fermion systems are typically compounds containing Ce or U or some other atom with an unfilled 4f or 5f shell, and they show extremely high susceptibilities χ as well as their high values of γ. We might indeed expect this from (2.3) and (2.6), which show that χ, like γ, is proportional to the density of states $g(\varepsilon_F)$ at the FS. What causes $g(\varepsilon_F)$ to be so high is not yet clear, but it is probably a many-body effect involving spin fluctuations. Whatever its nature, it seems to affect only electrons near ε_F, in much the same way that electron–phonon interactions affect only these electrons, because the shape of the FS in UPt_3, as determined experimentally, is very close to that predicted by band structure calculations.

Some, but not all, of the heavy-fermion materials become superconducting, though only at very low temperatures. Here again, the mechanism that produces the pairing is not clear, and this is another problem still awaiting solution.

In the following chapters, we shall largely ignore any possible complications due to many-body effects, and assume that their effect has been allowed for by appropriate choice of $\varepsilon(\boldsymbol{k})$.

7

The dynamics of Bloch electrons

7.1 THE VELOCITY AND THE LORENTZ FORCE

In Chapter 2 we were concerned only with free electrons, for which $m\boldsymbol{v} = \hbar \boldsymbol{k}$, and we assumed that the effect of fields $\boldsymbol{E}$ and $\boldsymbol{B}$ on an electron moving with velocity $\boldsymbol{v}$ was given by the Lorentz force equation

$$m\dot{\boldsymbol{v}} = \hbar\dot{\boldsymbol{k}} = -e(\boldsymbol{E} + \boldsymbol{v} \times \boldsymbol{B}) \tag{2.16}$$

For Bloch electrons, we no longer have $m\boldsymbol{v} = \hbar\boldsymbol{k}$. The electron experiences a large and rapidly fluctuating force as it moves through the potential $V(\boldsymbol{r})$, so that its instantaneous velocity $\boldsymbol{v}_{\mathrm{inst}}$ and momentum $m\boldsymbol{v}_{\mathrm{inst}}$ likewise fluctuate rapidly. However, its mean velocity, averaged over a unit cell, is well-defined, and is given by the usual quantum-mechanical expression

$$\boldsymbol{v} = -(i\hbar/2m)\int(\psi^*\nabla\psi - \psi\nabla\psi^*)\,\mathrm{d}^3\boldsymbol{r}$$

Applied to Bloch electrons, this expression leads eventually to a surprisingly simple result: for an electron in state $\boldsymbol{k}$, with energy $\varepsilon_n(\boldsymbol{k})$,

$$\boldsymbol{v}_n(\boldsymbol{k}) = \frac{1}{\hbar}\frac{\mathrm{d}\varepsilon_n(\boldsymbol{k})}{\mathrm{d}\boldsymbol{k}} \tag{7.1}$$

where $\mathrm{d}\varepsilon_n(\boldsymbol{k})/\mathrm{d}\boldsymbol{k} = \nabla_{\boldsymbol{k}}\varepsilon_n(\boldsymbol{k})$ is a vector with components $\partial\varepsilon_n/\partial k_x$, $\partial\varepsilon_n/\partial k_y$, $\partial\varepsilon_n/\partial k_z$. The derivation of (7.1) is straightforward but rather lengthy, and we shall not go through it here, but we can see that the result is plausible: if we think of the Bloch wave as having a time-dependence $e^{-i\omega t}$, where $\hbar\omega = \varepsilon_n(\boldsymbol{k})$, then a wave-packet built up of a small group of Bloch waves can be expected to move with the group velocity $\boldsymbol{v}_g = \mathrm{d}\omega/\mathrm{d}\boldsymbol{k}$, and this at once yields (7.1).

Since Bloch states are eigenstates, it follows that an electron in a Bloch state (or a wave-packet formed from Bloch states) will stay in that state, travelling with the velocity $\boldsymbol{v}$, indefinitely. We saw in section 2.3 that in pure metals at low temperatures the mean free path could become very large, and we now see how this comes about. In a perfect crystal, the mean free path is infinite. Only if the potential $V(\boldsymbol{r})$ is not perfectly periodic, so that the Bloch state is no longer an exact eigenstate, will the electron eventually be scattered out of it.

It follows from (7.1) that $\boldsymbol{v}(\boldsymbol{k})$ at any point $\boldsymbol{k}$ is normal to the constant-energy surface through $\boldsymbol{k}$ (just as, for example, an electric field $\boldsymbol{E}(\boldsymbol{r}) = -\nabla\phi(\boldsymbol{r})$ is normal to the equipotential surface $\phi(\boldsymbol{r})$). If the separation between two constant-energy surfaces ε and $\varepsilon + \delta\varepsilon$ varies, as shown in Fig. 7.1, $v_{\boldsymbol{k}} = |\boldsymbol{v}_{\boldsymbol{k}}|$ will be large where the normal separation $\delta k_\perp$ is small, and vice versa. In fact the volume $\delta V_{\boldsymbol{k}}$ enclosed between two adjacent energy surfaces can be written in terms of an integral over the $\boldsymbol{k}$-space surface area $S_{\boldsymbol{k}}$:

$$\delta V_{\boldsymbol{k}} = \int \delta k_\perp \,\mathrm{d}S_{\boldsymbol{k}} = \delta\varepsilon \int \mathrm{d}S_{\boldsymbol{k}}/\hbar v_{\boldsymbol{k}}$$

so that from (1.23) the density of states can be written

$$g(\varepsilon) = \int \mathrm{d}S_{\boldsymbol{k}}/4\pi^3\hbar v_{\boldsymbol{k}} \qquad (7.2)$$

If states of energy ε exist in more than one band, (7.2) will yield the total density of states if the integral is taken over all surfaces of energy ε in all bands, or if the integral is confined to the surfaces in band n it will yield $g_n(\varepsilon)$.

It follows from (7.2) that $g_n(\varepsilon)$ will be large in bands in which $v_{\boldsymbol{k}}$ is small, and vice versa. We saw in section 5.3 that $g_n(\varepsilon)$ must be large in narrow bands, where $\Delta\varepsilon = \varepsilon_{\max} - \varepsilon_{\min}$ is small, and (7.2) is consistent with this; (7.1) shows that the average speed $\bar{v}$ must be small in narrow bands, since $\bar{v} \sim \Delta\varepsilon/\hbar K$, where $2K$ is the distance across the BZ.

When a Bloch electron is subjected to a Lorentz force $\boldsymbol{F} = -e(\boldsymbol{E} + \boldsymbol{v} \times \boldsymbol{B})$ due to externally applied fields $\boldsymbol{E}$ and $\boldsymbol{B}$,

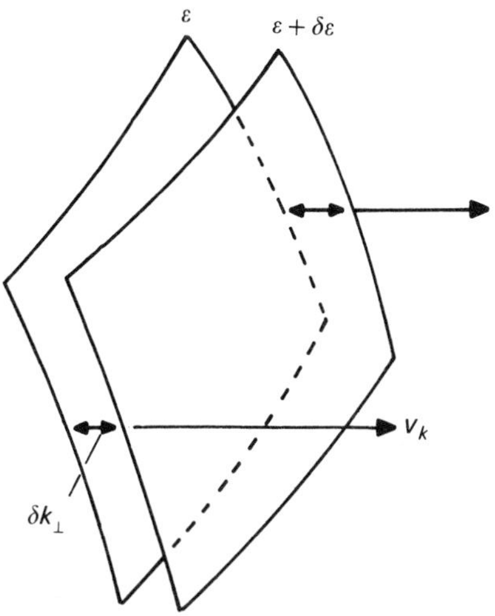

Fig. 7.1 The separation $\delta k_\perp$ between the two adjacent energy surfaces is smaller on the near side than on the far side, and $\mathbf{v}_k$ is therefore larger on the near side.

it responds as though its momentum were $\hbar\boldsymbol{k}$; that is, the value of $\hbar\boldsymbol{k}$ changes at a rate

$$\hbar\dot{\boldsymbol{k}} = \boldsymbol{F} = -e(\boldsymbol{E} + \boldsymbol{v} \times \boldsymbol{B}) \tag{7.3}$$

(with $\boldsymbol{v}$ given by (7.1)). The quantity $\hbar\boldsymbol{k}$ is sometimes called the 'crystal momentum' of the electron to distinguish it from the actual momentum, whose mean value is $m\boldsymbol{v}$; only for free electrons are the two equal.

We cannot write $m\dot{\boldsymbol{v}} = \boldsymbol{F}$ for a Bloch electron because when it is accelerated by an external force $\boldsymbol{F}$ it also experiences a reaction force $\boldsymbol{F}_{\mathrm{r}}$ due to the lattice. Because of this, the momentum $m\boldsymbol{v}$ is not a particularly useful quantity for Bloch electrons. To see what effect $\boldsymbol{F}$ does have on $\boldsymbol{v}$, we need to relate $\dot{\boldsymbol{v}}$ to $\dot{\boldsymbol{k}}$. We can write

$$\begin{aligned}\dot{v}_i &= \sum_j (\partial v_i/\partial k_j)\dot{k}_j \qquad \text{(where } i, j = x, y, \text{ or } z)\\ &= \hbar^{-1}\sum_j (\partial^2\varepsilon/\partial k_i\partial k_j)\dot{k}_j\\ &= \hbar^{-2}\sum_j (\partial^2\varepsilon/\partial k_i\partial k_j)\dot{F}_j\end{aligned}$$

since $\dot{k}_j = \hbar^{-1} F_j$, from (7.3). Comparing this with the result $\dot{\boldsymbol{v}} = m^{-1}\boldsymbol{F}$ for free electrons, we see that a Bloch electron behaves as though m^{-1} was a *tensor* – the 'reciprocal mass tensor' – with components

$$(m^{-1})^*_{ij} = \hbar^{-2}(\partial^2 \varepsilon/\partial k_i \partial k_j) \tag{7.4}$$

This tensor, again, is usually not a particularly useful quantity in metals, but it is useful in semiconductors, where the energy surfaces are to a good approximation ellipsoidal, so that (if the x, y, z axes are chosen appropriately) we can write $\varepsilon(\boldsymbol{k}) \approx \alpha_x k_x^2 + \alpha_y k_y^2 + \alpha_z k_z^2$. Then $(m^{-1})^*_{ij}$ vanishes unless $i = j$, and we are left with the three components $(m^{-1})^*_{ii} = 2\alpha_i/\hbar^2$.

It is quite difficult to prove (7.3), and we shall not try to do so here. One can make it look plausible by arguing that the change of energy $\delta\varepsilon$ of the electron in time δt must equal the work done by the force $\boldsymbol{F}$: $\delta\varepsilon = \boldsymbol{F}\cdot\boldsymbol{v}\,\partial t$. Writing $\delta\varepsilon = (\mathrm{d}\varepsilon/\mathrm{d}\boldsymbol{k})\cdot\delta\boldsymbol{k} = \hbar\boldsymbol{v}\cdot\delta\boldsymbol{k}$, this yields

$$h\boldsymbol{v}\cdot\dot{\boldsymbol{k}} = \boldsymbol{F}\cdot\boldsymbol{v} \tag{7.5}$$

If we then spuriously 'cancel out' the $\boldsymbol{v}$ on each side, we arrive at (7.3). But all that (7.5) really tells us is that $h\dot{\boldsymbol{k}}_{\parallel} = \boldsymbol{F}_{\parallel}$, where $\boldsymbol{k}_{\parallel}$ and $\boldsymbol{F}_{\parallel}$ are the components of $\boldsymbol{k}$ and $\boldsymbol{F}$ parallel to $\boldsymbol{v}$. In particular, (7.5) tells us very little about the effect of $\boldsymbol{B}$, because the force $-e\boldsymbol{v}\times\boldsymbol{B}$ has no component parallel to $\boldsymbol{v}$, so that $\boldsymbol{F}\cdot\boldsymbol{v} = 0$. Nevertheless, (7.3) can be proved true, and is basic to the theory of the electrical and magnetic properties of conductors.

If $\boldsymbol{B} = 0$, (7.3) shows that the effect of $\boldsymbol{E}$ is to move the whole electron assembly through $\boldsymbol{k}$-space at the uniform rate $-e\boldsymbol{E}/\hbar$. For a full band, this displacement will produce no resultant current, because the band remains just as full as before; in the reduced zone scheme, the electrons which pass out through one face of the BZ immediately reappear, unchanged except for a relabelling of their $\boldsymbol{k}$ vector, at the opposite face. We still have $\sum\boldsymbol{v} = 0$, summed over all the electrons in the band, and therefore $\boldsymbol{J} = 0$. Thus in a material containing only completely filled or completely empty bands, no current can flow, and band theory thus accounts at once for the existence of insulators. (In a very strong electric field, electrons may be able to tunnel through the band gap out of a filled band into an empty one, leading to dielectric breakdown. In a semiconducting p–n junction, similar tunnelling can occur across the junction; this sets in very suddenly at a particular bias voltage and forms the basis of the Zener diode.)

In metals, with partly-filled bands, the displacement of the FS by $\boldsymbol{E}$ would very rapidly lead to a gigantic current flow, if it were not for collisions. Collisions effectively limit the displacement of the FS to an amount $\boldsymbol{k}_\mathrm{d} = -e\boldsymbol{E}\tau/\hbar$, which in all practical circumstances will be extremely small compared with the size of the BZ (problem 7.1). We shall consider the effect of $\boldsymbol{E}$ in more detail in Chapter 9; for the rest of the present chapter we shall assume $\boldsymbol{E} = 0$, and consider just the effect of a magnetic field $\boldsymbol{B}$.

7.2 ORBITS IN A MAGNETIC FIELD

As equation (7.3) shows, the effect of $\boldsymbol{B}$ is quite different from the effect of $\boldsymbol{E}$. First, $\dot{\boldsymbol{k}}$ is at right angles to $\boldsymbol{B}$, so that the component k_B parallel to $\boldsymbol{B}$ stays constant; $\boldsymbol{k}$ moves through $\boldsymbol{k}$-space in a plane normal to $\boldsymbol{B}$. Secondly, $\dot{\boldsymbol{k}}$ is at right angles to $\boldsymbol{v}$; but $\boldsymbol{v}$ itself is normal to the constant-energy surface through the point $\boldsymbol{k}$, so that $\dot{\boldsymbol{k}}$ must lie *in* that surface. In other words, because the magnetic field does no work on the electron, $\boldsymbol{k}$ must stay on one constant-energy surface as it moves. These two facts, taken together, define the orbit in $\boldsymbol{k}$-space completely, for a given starting-point $\boldsymbol{k}_0$: if we imagine the constant-energy surface $\varepsilon = \varepsilon(\boldsymbol{k}_0)$ sliced through by a plane normal to $\boldsymbol{B}$ which passes through the point $\boldsymbol{k}_0$, the curve in which the plane intersects the surface will be the electron orbit.

For free electrons, where the constant-energy surfaces are spheres, the resultant orbits are obviously circles, as we would expect from section 2.4. But for more complex energy surfaces, the orbits can be of several different kinds. Consider a simple cubic metal, with the FS shown in Fig. 7.2a. In the periodic zone scheme, this FS will become a three-dimensional network, as shown in Fig. 7.2b. This shows two possible slice-planes normal to $\boldsymbol{B}$, for $\boldsymbol{B}$ along [100]. The closed orbits on the left enclose filled states, and are qualitatively no different from free-electron orbits; but those on the right enclose empty states, and are called 'hole' orbits. On these orbits, $\boldsymbol{v}$ is directed inwards instead of outwards, and this means that if $\boldsymbol{k}$ moves clockwise round the electron orbits on the left, it must move anticlockwise round the hole orbits on the right, as shown. As we shall see later, it is these hole orbits which enable us to understand the existence of positive Hall coefficients, which were inexplicable on the free-electron model.

Closed energy surfaces can be either electron or hole surfaces, depending on whether the filled states are inside or outside. The second-band surface in Fig. 5.4 is an example of a hole surface, and for any direction of $\boldsymbol{B}$ it will give rise only to hole orbits. The surface shown in Fig. 7.2 is an 'open' surface, extending through the whole of $\boldsymbol{k}$-space in the periodic zone scheme, and cannot at first sight be classified as either an electron or a hole surface; as Fig. 7.2b shows, it can give rise simultaneously to both electron orbits and hole orbits.

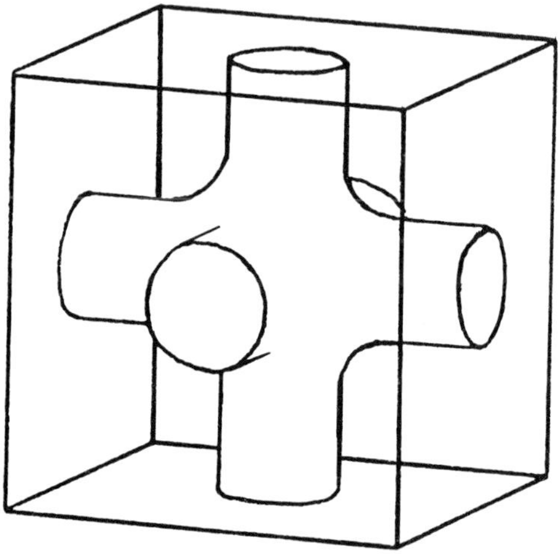

(a)

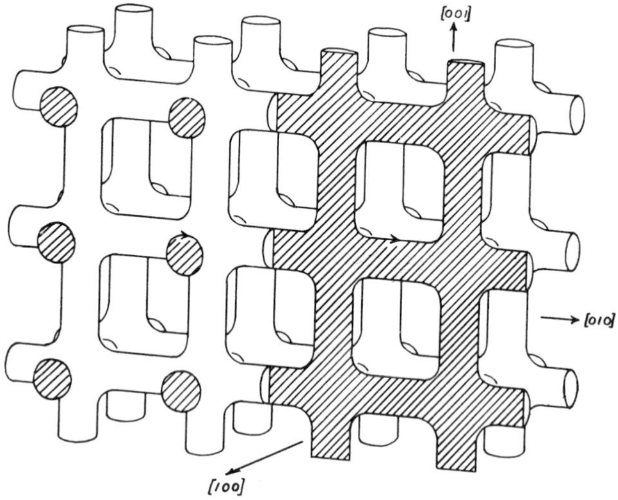

(b)

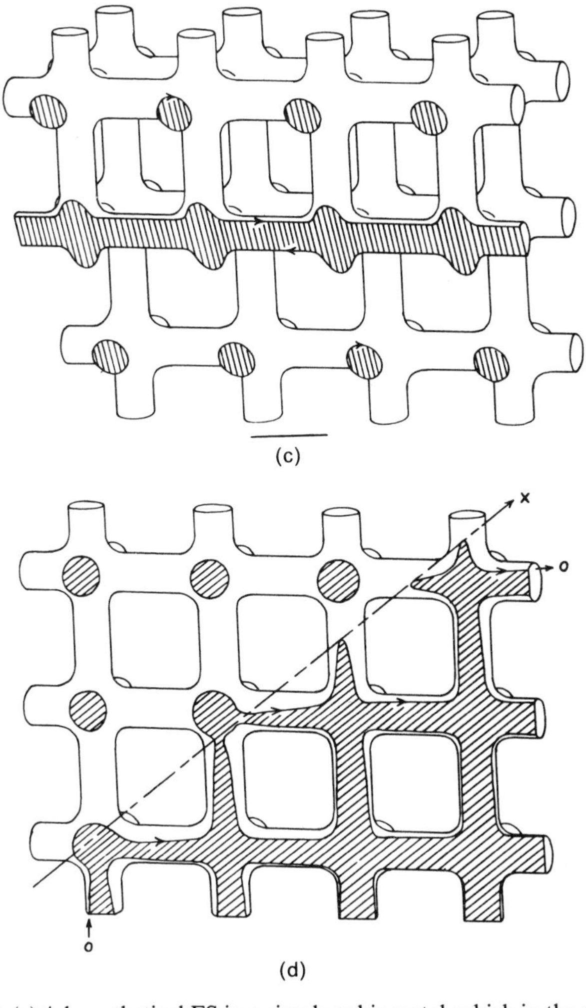

Fig. 7.2 (a) A hypothetical FS in a simple cubic metal, which in the periodic zone scheme gives rise to the network shown in (b–d). (b) Closed electron and hole orbits. (c) Periodic open orbits. (d) An aperiodic open orbit. [From Chambers (1960), in The Fermi Surface (eds Harrision and Webb), Wiley, Chichester, p. 100.]

But there will always be field directions $\boldsymbol{B}$ which generate only closed orbits on such a surface, and the nature of the predominant orbits then determines whether it is an electron surface or a hole surface. In Fig. 7.2, we clearly have an electron surface.

If $\boldsymbol{B}$ is tilted slightly away from the [100] direction towards [001], the slice-planes will be correspondingly tilted, as shown in Fig. 7.2c, and will now generate a pair of *open orbits*, as shown, which extend through the whole of $\boldsymbol{k}$-space. More generally, if $\boldsymbol{B}$ is tilted in some arbitrary direction, as shown in Fig. 7.2d, a single slice-plane may generate electron orbits in one region and hole orbits in another. It must then generate an open orbit OO, as shown, as the 'coast-line' between the two regions. This orbit again, like open orbits generally and unlike closed orbits, extends through the whole of $\boldsymbol{k}$-space. Although its path may be tortuous in detail, it must run on average in a well-defined direction, along the line in which the slice-plane intersects an adjacent symmetry plane in $\boldsymbol{k}$-space – the (100) plane, in this example. Open orbits, when they exist, have an important effect on the conduction properties of a metal in a magnetic field.

Although power dissipation limits $\boldsymbol{E}$ to very small values in good conductors (problem 7.1), there is no such limit to $\boldsymbol{B}$, because the electrons gain no energy from $\boldsymbol{B}$. Correspondingly, although attainable fields $\boldsymbol{E}$ produce only extremely small displacements in $\boldsymbol{k}$-space before collisions occur, attainable fields $\boldsymbol{B}$ can produce very large displacements – large enough, in a pure metal where τ is large, for the $\boldsymbol{k}$-vector to complete several circuits around a closed orbit, or to move a long way through the periodic zone scheme in an open orbit, before collision.

For closed orbits, we can derive a useful expression for the cyclotron period τ_c – the time taken for the electron to travel once around the orbit – or for the cyclotron frequency $\omega_c = 2\pi/\tau_c$. We can write (7.3) as $\dot{k}_t = ev_n\boldsymbol{B}/\hbar$, where k_t is measured around the orbit and $\boldsymbol{v}_n$ is the component of $\boldsymbol{v}$ normal to $\boldsymbol{B}$. We can also write $v_n = \hbar^{-1}\mathrm{d}\varepsilon/\mathrm{d}k_n$, where k_n is measured normal to the orbit, in the slice-plane (Fig. 7.3). Writing $\dot{k}_t = \delta k_t/\delta t$, we then have

$$\delta t = \delta k_t/\dot{k}_t = \delta k_t \delta k_n \hbar^2/e\boldsymbol{B}\delta\varepsilon$$

Now $\delta k_t \delta k_n$ is an element of area in $\boldsymbol{k}$-space between the orbits of energy ε and energy $\varepsilon + \delta\varepsilon$, as shown in Fig. 7.3, so that if we integrate around the whole orbit we have

$$\tau_c = (\hbar^2/e\boldsymbol{B})(\partial\mathscr{A}/\partial\varepsilon)_{k_B}$$

where $\mathscr{A}$ is the orbit area in $\boldsymbol{k}$-space. We can thus write the cyclotron

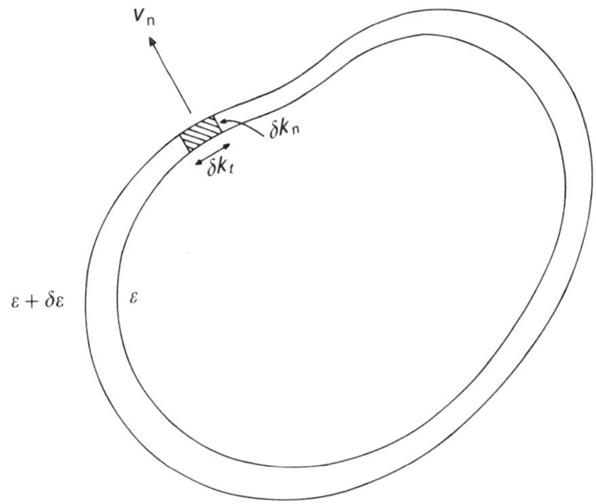

Fig. 7.3 The area $\delta\mathscr{A}$ enclosed between two adjacent orbits on a given slice-plane is given by $\delta\mathscr{A} = \oint \delta k_n \, \mathrm{d}k_t$.

frequency as

$$\omega_c = eB/m_c^* \tag{7.6}$$

where the 'cyclotron mass' m_c^* is given by

$$m_c^* = (\hbar^2/2\pi)(\partial\mathscr{A}/\partial\varepsilon)_{k_B} \tag{7.7}$$

(Note that for a hole orbit, $\mathscr{A}$ decreases as ε increases, so that m_c^* is negative; the changed sign of m_c^* and hence of ω_c reflects the different sense of rotation around electron and hole orbits.)

As the electron traverses its orbit in $\boldsymbol{k}$-space, it will at the same time (if we think of it as a localized wave-packet) pursue a corresponding path through real space. We can find the form of this path from (7.3): since $\boldsymbol{v} = \dot{\boldsymbol{r}}$, we have (for $\boldsymbol{E} = 0$)

$$\dot{\boldsymbol{k}} = -(e/\hbar)\dot{\boldsymbol{r}} \times \boldsymbol{B}$$

so that

$$\boldsymbol{k} - \boldsymbol{k}_B = -(e/\hbar)(\boldsymbol{r} - \boldsymbol{r}_0) \times \boldsymbol{B} \tag{7.8}$$

where as before $\boldsymbol{k}_B$ is the component of $\boldsymbol{k}$ parallel to $\boldsymbol{B}$, and $\boldsymbol{r}_0$ is an integration constant which determines the position of the real-space orbit in the metal. If we choose the direction of $\boldsymbol{B}$ as the z axis, (7.8) becomes

$$x - x_0 = (\hbar/eB)k_y; \quad y - y_0 = -(\hbar/eB)k_x \tag{7.9}$$

This result applies to both closed and open orbits. It tells us that the real-space motion, projected on the xy plane normal to $\boldsymbol{B}$, is an orbit of exactly the same shape as the $\boldsymbol{k}$-space orbit, but rotated through 90° and scaled by the factor $\hbar/eB$. The electron will normally also have a component of velocity in the direction of $\boldsymbol{B}$ (and except for free electrons this component will vary in magnitude as the electron traverses the orbit), so that if the $\boldsymbol{k}$-space orbit is closed, the real-space orbit will be a distorted helix, with its axis in the $\boldsymbol{B}$ direction, along z, and its shape in the xy plane given by (7.9). The 'pitch' of the helix will be $\bar{v}_z \tau_c$, where $\bar{v}_z$ is the average value of v_z around the orbit. If the $\boldsymbol{k}$-space orbit is open, and runs say in the k_x direction, (7.9) shows that the xy projection of the real-space orbit will run in the y direction.

In many metals, the FS is an open surface; in Cu, for example, the FS contacts the BZ boundary on the [111] faces (Fig. 5.7) to form a complex three-dimensional network in the periodic zone scheme, which can support a variety of electron orbits, hole orbits and open orbits. Open energy surfaces, like van Hove singularities, necessarily occur somewhere between the minimum and maximum energies ε_{min} and ε_{max} as a band is filled up. When the band first starts to fill, at energies just above ε_{min}, the filled states will form small isolated pockets, and the energy surfaces will be closed electron surfaces. When the band is almost full, at energies just less than ε_{max}, the remaining empty states will form small isolated pockets – closed hole surfaces. In between, open surfaces must occur.

In semiconductors, by definition, all bands are either almost completely empty or almost completely full, and the relevant energy surfaces are either electron surfaces or hole surfaces; open surfaces play no part in the properties of semiconductors.

7.3 ORBIT QUANTIZATION

In the last two sections, we have treated the electron as a wave-packet built up of a small group of Bloch waves, reasonably well localized in both $\boldsymbol{k}$-space and real space and moving through $\boldsymbol{k}$-space and real space according to the equations of motion (7.3) and (7.8). For most purposes this 'semi-classical' approach is entirely adequate, but at one point we have to be a little more careful; although (7.3) is correct, it does not tell the whole story. As it stands, our rule for tracing out an electron orbit in $\boldsymbol{k}$-space, as the intersection of a slice-plane $\boldsymbol{k}_B = \text{constant}$ with a constant-energy surface, allows a continuous spectrum of orbit energies – if an orbit of energy ε can exist on a given slice-plane, so can an orbit of energy $\varepsilon + \delta\varepsilon$, with $\delta\varepsilon$ as small as we like. Now for open orbits this conclusion is in fact correct, but not otherwise; if the $\boldsymbol{k}$-space orbits are closed, their energies turn out to be quantized, as indeed we might have expected intuitively.

For free electrons, it is not too difficult to solve the Schrödinger equation in a magnetic field B_z, and it turns out that the energy separation between adjacent quantized orbits, for given k_z, is just $\hbar\omega_c$ where ω_c is the cyclotron frequency eB_z/m (cf. (7.6)). The allowed energies are

$$\varepsilon = \hbar^2 k_z^2/2m + (n + \tfrac{1}{2})\hbar\omega_c \tag{7.10}$$

where the first term represents the kinetic energy of motion along the field direction.

For Bloch electrons in a magnetic field, the Schrödinger equation is a good deal more difficult to solve, and the appropriate quantization rule for Bloch electrons was first derived by Onsager in 1952 in a different way, using Bohr's 'old quantum theory' of 1913. On this theory, which predated wave mechanics, the motion of a particle in a closed orbit had to satisfy the Bohr–Sommerfeld quantization rule:

$$\oint \boldsymbol{p}\cdot \mathrm{d}\boldsymbol{r} = 2\pi(n + \gamma)\hbar \tag{7.11}$$

where n is an integer, γ is a constant, $\boldsymbol{p}$ is the momentum variable 'conjugate' to $\boldsymbol{r}$ and the integration is around the closed orbit. For a Bloch electron in a magnetic field $\boldsymbol{B}$, the appropriate form to choose for $\boldsymbol{p}$ is $\boldsymbol{p} = \hbar\boldsymbol{k} - e\boldsymbol{A}$ (Appendix A), where $\boldsymbol{A}$ is the vector potential, so that curl $\boldsymbol{A} = \boldsymbol{B}$.

With this choice for $\boldsymbol{p}$, (7.11) becomes

$$\oint (\hbar\boldsymbol{k} - e\boldsymbol{A})\cdot \mathrm{d}\boldsymbol{r} = 2\pi(n + \gamma)\hbar \tag{7.12}$$

where the integration is taken around the closed orbit in the xy plane, normal to $\boldsymbol{B}$, defined by (7.9). It is not difficult to evaluate the integral (problem 7.3), and hence to show that (7.12) leads to Onsager's result

$$\mathscr{A}_{\mathrm{r}}B = 2\pi(n+\gamma)\hbar/e \tag{7.13}$$

where $\mathscr{A}_{\mathrm{r}}$ is the area enclosed by the real-space orbit, or more precisely by its projection on the xy plane. This equation tells us that the allowed orbits are those for which the enclosed flux $\mathscr{A}_{\mathrm{r}}B$ is (apart from γ) an integral number of 'flux units' $2\pi\hbar/e$ ($=4.136\times10^{-15}$ Wb, or $4.136\times10^{-7}\,\mathrm{G\,cm^2}$).

From equation (7.9), the area $\mathscr{A}$ of the $\boldsymbol{k}$-space orbit is related to $\mathscr{A}_{\mathrm{r}}$ by $\mathscr{A}_{\mathrm{r}}=(\hbar/eB)^2\mathscr{A}$, so that (7.13) can be rewritten in terms of $\mathscr{A}$: the allowed values of $\mathscr{A}$ are given by

$$\mathscr{A} = 2\pi(eB/\hbar)(n+\gamma) \tag{7.14}$$

Since, for given k_z, $\mathscr{A}$ is a function of energy ε, (7.14) tells us how the orbit energy is quantized.

Although Onsager derived (7.13) and (7.14) from the semi-classical quantization rule (7.12), solution of the Schrödinger equation for Bloch electrons leads to exactly the same results, and also shows that $\gamma=\frac{1}{2}$. For free electrons, (7.14) leads at once to the correct result (7.10) for the allowed energies (problem 7.4). For Bloch electrons, the allowed energies are no longer given by (7.10), but the energy separation $\delta\varepsilon$ between adjacent orbits, for given k_z, is still given by $\delta\varepsilon=\hbar\omega_{\mathrm{c}}$, with ω_{c} given by (7.6) and (7.7). To show this, we note first that except for very small orbit areas the quantum number n in (7.14) will typically be of order 10^4 or more (problem 7.5). We can thus assume that over a small range of n, ε will vary linearly with $\mathscr{A}$, and write $\delta\varepsilon=(\partial\varepsilon/\partial\mathscr{A})\delta\mathscr{A}=(\partial\varepsilon/\partial\mathscr{A})2\pi e\boldsymbol{B}/\hbar$. Using (7.6) and (7.7), we at once find that $\delta\varepsilon=\hbar\omega_{\mathrm{c}}$.

A given $\boldsymbol{k}$-space orbit in a field B_z is specified completely by k_z and n, but the corresponding real-space orbit is not. There are many different real-space orbits corresponding to a given $\boldsymbol{k}$-space orbit, all having the same energy and the same helical form, and all running parallel to one another in the direction of the z axis, but located around different points x, y in the plane normal to B_z. How many physically distinct real-space orbits are there, in a crystal of given size, for each $\boldsymbol{k}$-space orbit? In other words, what is the degeneracy

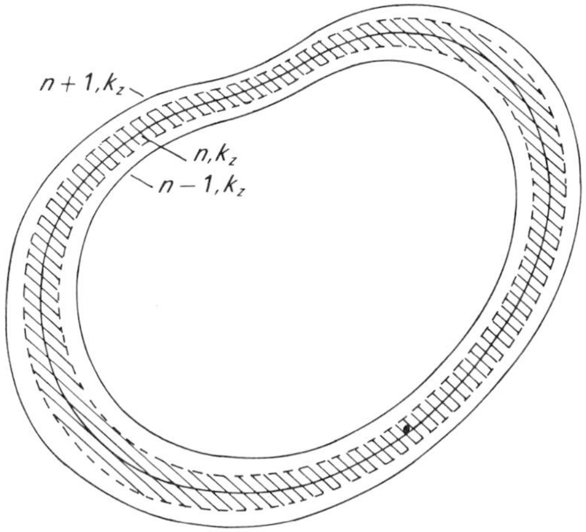

Fig. 7.4 The degeneracy of the orbit n, k_z is equal to the number of states which would, in $\boldsymbol{B}=0$, occupy the shaded area – the area closer to this orbit than to either of the two adjacent orbits.

of each $\boldsymbol{k}$-space orbit? The answer is easy enough to state, though difficult to prove, and we shall simply state it: the degeneracy is such that the average density of states in $\boldsymbol{k}$-space is the same as it was in $\boldsymbol{B}=0$.

To put that more precisely, consider a particular $\boldsymbol{k}$-space orbit (n, k_z). In the plane $k_z = \text{constant}$, there will be a region of area $\delta\mathscr{A} = 2\pi eB/\hbar$ which is closer to this orbit than to either of the adjacent orbits $(n+1, k_z)$ and $(n-1, k_z)$ as shown shaded in Fig. 7.4. Suppose that this shaded area contains N Bloch states when $\boldsymbol{B}=0$. Then the degeneracy of the orbit (n, k_z) is just N; we can think of all these N states as 'condensing' on to the orbit (n, k_z) when the field is applied. At the same time, the energies of these states, which range from $\varepsilon_n - \frac{1}{2}\hbar\omega_c$ to $\varepsilon_n + \frac{1}{2}\hbar\omega_c$ in $\boldsymbol{B}=0$, likewise condense on to the one highly degenerate energy level ε_n, as shown in Fig. 7.5a.

So far we have neglected entirely the effect of electron spin on the energy levels. As we saw in section 2.2, a field $\boldsymbol{B}$ will alter the energy of a spin-up or spin-down electron by an amount $\pm\mu_B B$, where $\mu_B = e\hbar/2m$ is the Bohr magneton. This result holds for free electrons; for electrons moving through a lattice, spin–orbit interactions may cause the electron to behave as though it had an 'effective g-value' g_{eff} different from the free-electron value $g=2$, and consequently as though it had an effective magnetic moment $\mu_{\text{eff}} = (g_{\text{eff}}/2)\mu_B$.

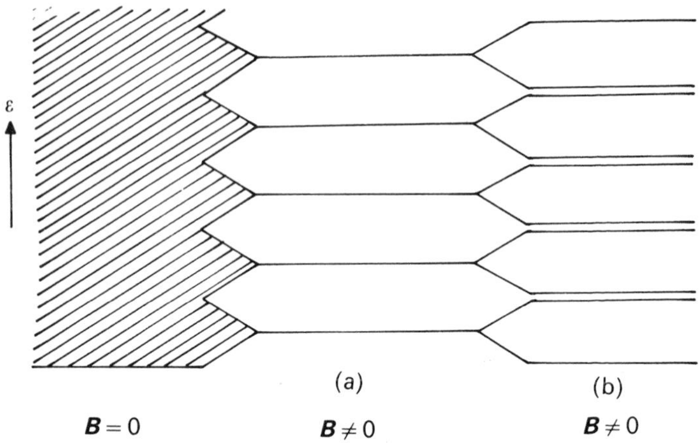

Fig. 7.5 Showing the continuum of energy levels in $\boldsymbol{B} = 0$, and the effect of (a) quantization of the orbital energy and (b) the further splitting of energy levels due to spin.

If we divide the quantized orbital states into two sub-groups, the spin-up states and the spin-down states, the effect of spin will thus be to displace the whole energy spectrum of one sub-group upwards by an amount $\mu_{\text{eff}}B$, and that of the other sub-group downwards by the same amount, so that the total spin splitting is $2\mu_{\text{eff}}B$. For free electrons, $2\mu_{\text{eff}}B = 2\mu_{\text{B}}B = e\hbar B/m$, which is exactly equal to the separation of orbital levels $\hbar\omega_{\text{c}}$, and the net effect is that the spin-down states in orbital level $n+1$ now coincide in energy with the spin-up states in level n, as shown in Fig. 7.5b. For Bloch electrons, the displaced levels will not coincide in this way, because we then have $2\mu_{\text{eff}}B = (g_{\text{eff}}/2)e\hbar B/m$, whereas $\hbar\omega_{\text{c}} = e\hbar B/m_{\text{c}}^*$, and there is no reason why these two quantities should be equal.

We saw in section 2.2 how the electron spin leads to Pauli paramagnetism. In 1930 Landau showed that the orbital quantization leads to a slight increase in the total energy U of the electron assembly, and therefore in the free energy $F = U - TS$, and hence leads to an additional *diamagnetic* term: the 'Landau diamagnetism' χ_{L}. (According to statistical mechanics, the magnetic moment M of any body in a field B is given by $M = -\partial F/\partial B$, where F is the free energy of the body. The susceptibility is thus given by

$\chi = \partial M/\partial B = -\partial^2 F/\partial B^2$.) For a free-electron metal, as Landau showed,

$$\chi_L = -e^2 k_F/12\pi^2 m \tag{7.15}$$

This diamagnetic term, arising from the quantized orbital motion of the electrons (Fig. 7.5a), is quite separate from the Pauli paramagnetism χ_P arising from the spin splitting (Fig. 7.5b). As shown in section 2.2,

$$\chi_P = \mu_B^2 g(\varepsilon_F) \tag{2.6}$$

where $g(\varepsilon_F)$ is the density of states at ε_F. Using (1.26), we can write $g(\varepsilon_F) = mk_F/\pi^2\hbar^2$ for free electrons, so that (with $\mu_B = e\hbar/2m$) (2.6) becomes

$$\chi_P = e^2 k_F/4\pi^2 m \tag{7.16}$$

and thus $\chi_L = -\chi_P/3$. This result holds only for free electrons; in a periodic lattice, (7.15) must be replaced by a much more complicated expression, involving an integral over the FS, and in (2.6) we must replace μ_B by μ_{eff} (which may also vary over the FS) so that in general there is no reason to expect $\chi_L = -\chi_P/3$.

7.4 THE DE HAAS–VAN ALPHEN EFFECT

As we noted in section 2.2, it is difficult to sort out experimentally the various contributions to the steady susceptibility χ. Of much more interest is another effect of orbit quantization, also predicted by Landau in 1930 and independently discovered experimentally in the same year by de Haas and van Alphen: an *oscillatory* variation of M and χ with B (or more precisely with $1/B$).

To see how this arises, consider what the allowed energy surfaces look like in $\boldsymbol{k}$-space. In any plane of constant k_z, the constant-energy contours may form either closed or open orbits. The open orbits (if any) do not concern us here, because they are not quantized. The closed orbits must satisfy the quantization condition (7.14) (with $\gamma = \frac{1}{2}$), which says that the nth orbit must have area $\mathscr{A}_n = 2\pi(eB/\hbar)(n + \frac{1}{2})$, independent of k_z. Thus the sequence of orbits in successive slice-planes k_z will form a tube, of constant cross-sectional area $\mathscr{A}_n$, running through $\boldsymbol{k}$-space in the general direction of the field B_z. For free electrons, these 'Landau tubes' are just circular cylinders, as shown in Fig. 7.6.

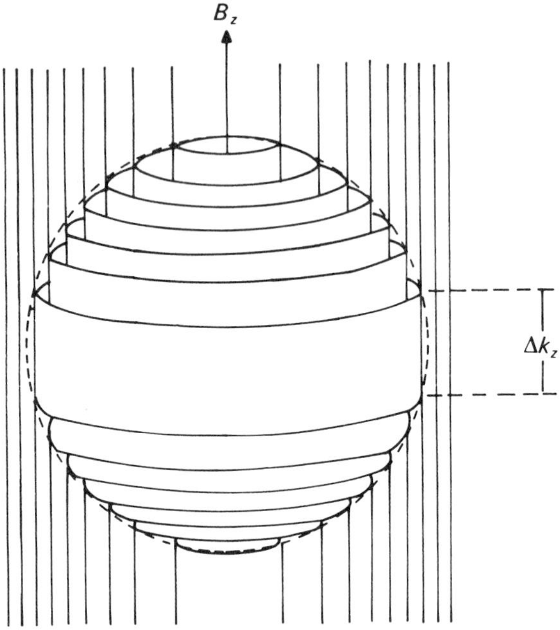

Fig. 7.6 Showing the origin of the dHvA effect. [From Chamber (1956), *Can J. Phys*, **34**, 1395.]

At $T = 0$, all states along these tubes will be completely filled up to $\varepsilon = \varepsilon_{F,0}$, and all higher states will be completely empty. Each tube can hold just the same number of electrons as the adjacent region of $\boldsymbol{k}$-space held in $\boldsymbol{B} = 0$, so that the average density of states in $\boldsymbol{k}$-space is the same as it was in $\boldsymbol{B} = 0$. The value of $\varepsilon_{F,0}$ will therefore be virtually identical with the $\boldsymbol{B} = 0$ value, so that the filled regions of the tubes lie within an FS which is virtually identical with the $\boldsymbol{B} = 0$ FS. For free electrons, this will be the spherical FS shown dashed in Fig. 7.6.

If B_z is increased slightly, the cross-sectional area $\mathscr{A}_n$ of each tube will likewise increase. For most of the tubes, this will lead simply to a small decrease in the occupied length of the tube Δk_z. But for the outermost occupied tube, the decrease will be much larger; it is clear from Fig. 7.6 that for this tube Δk_z will fall very rapidly to zero as the tube area $\mathscr{A}_n$ approaches the extremal cross-sectional area $\mathscr{A}_e$ of the FS (problem 7.6). This leads to a significant fluctuation in the total energy U of the electron assembly each time a tube passes out through the FS, and hence to fluctuations in M and χ – in other words, to the de Haas–van Alphen (dHvA) effect.

A fluctuation will occur every time $\mathscr{A}_n = \mathscr{A}_e$; that is, every time B satisfies the equation $\mathscr{A}_e = 2\pi(eB/\hbar)(n+\frac{1}{2})$, or

$$1/B = (2\pi e/\hbar\mathscr{A}_e)(n+\tfrac{1}{2}) \tag{7.17}$$

The fluctuations are therefore periodic in $1/B$, with a period $\Delta(1/B) = 2\pi e/\hbar\mathscr{A}_e$, and the great importance of the dHvA effect is that it enables us, by measuring this period, to deduce the extremal cross-sectional area $\mathscr{A}_e$ of the FS in a plane normal to $\boldsymbol{B}$. By making such measurements for a variety of directions of $\boldsymbol{B}$ relative to the crystal axes, we can learn a great deal about the size and shape of the FS, and indeed most of our experimental knowledge of the FS of different metals comes from such measurements.

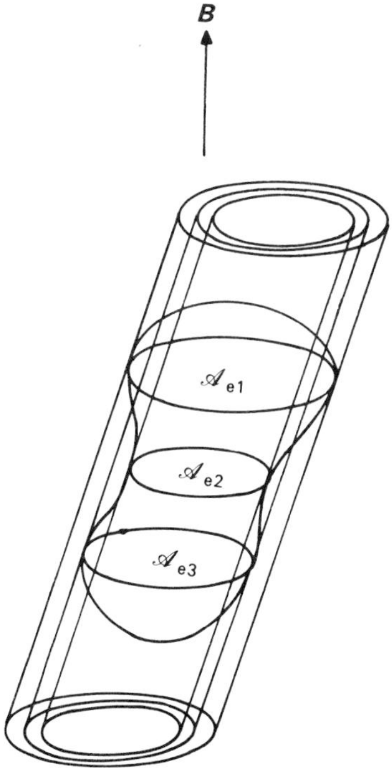

Fig. 7.7 An FS shaped like an hour-glass can show several extremal areas.

It is quite possible for an FS to have more than one extremal area for a given $\boldsymbol{B}$ direction. Figure 7.7, for example, shows an FS shaped like an hour-glass with its long axis at an angle to $\boldsymbol{B}$, so that the Landau tubes themselves run at an angle to $\boldsymbol{B}$. For this FS and this $\boldsymbol{B}$ direction there are three extremal areas: maxima at $\mathscr{A}_{e1}$, $\mathscr{A}_{e3}$ and a minimum at $\mathscr{A}_{e2}$. There will be a fluctuation in U and in M whenever one of the Landau tubes passes out of the FS through any one of these extrema; that is, whenever $\mathscr{A}_n = \mathscr{A}_{e1}$, $\mathscr{A}_{e2}$ or $\mathscr{A}_{e3}$. Correspondingly, M will now show three superposed periodicities as a function of $1/B$, and the experimental data will need careful analysis to disentangle these. Figure 7.8 shows an example of the complex patterns observed experimentally. Because the quantum numbers n are usually large, the oscillations can be observed over many cycles as B is varied; but for this, it would not be possible to interpret such complex patterns.

A full theoretical analysis of the dHvA effect would need several pages of detailed mathematics, and we shall not attempt it here, but it is worth looking at the resulting expression for M, because each term in this rather formidable-looking expression has a simple physical interpretation. Slightly simplified, the expression is

$$M = CFTB^{-1/2}|\mathscr{A}''_e|^{-1/2}\sin\left[2\pi\left(\frac{F}{B}-\frac{1}{2}\right)\mp\frac{\pi}{4}\right]\cos\left(\frac{\pi g_{\text{eff}}m_c^*}{2m}\right)$$
$$\times \exp-(2\pi^2 kT/\hbar\omega_c)\exp-(\pi/\omega_c\tau) \qquad (7.18)$$

Here C is a constant involving e, k_b and $\hbar$, $F=(\hbar/2\pi e)\mathscr{A}_e$, and $\mathscr{A}''_e = \partial^2\mathscr{A}/\partial k_z^2$, evaluated at $\mathscr{A}=\mathscr{A}_e$, where $\mathscr{A}$ is the cross-sectional area (csa) of the FS on the plane k_z. To make it easier to refer to the various terms, we can rewrite this expression as

$$M = CFTB^{-1/2}|\mathscr{A}''_e|^{-1/2}\sin\theta\cos\phi\exp-\alpha(T+T_D) \quad (7.18a)$$

where $kT_D = \hbar/2\pi\tau$.

If the FS has more than one extremal area, as in Fig. 7.7, each one will give a contribution to M of this form. (The oscillations of M are in fact not purely sinusoidal; there are also harmonic terms of the form $\sin(r\theta)\cos(r\phi)\exp - r\alpha(T+T_D)$, where $r = 2, 3, \ldots$, but these are usually much smaller, because of the extra r in the exponential factor.)

The most important term in (7.18) or (7.18a) is the $\sin\theta$ term: this

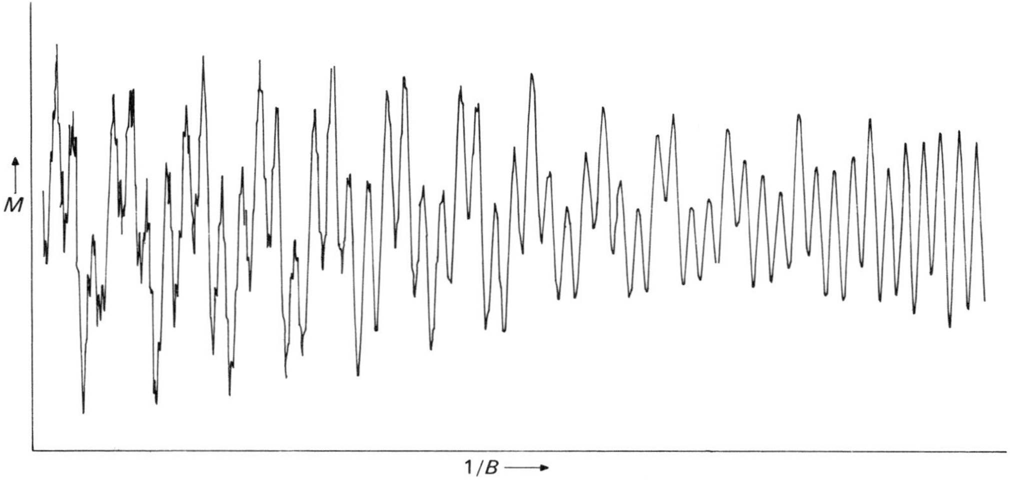

Fig. 7.8 The variation of M with $1/B$ for a particular field direction in Mg, showing several different oscillation frequencies. (Parsons, D., unpublished).

is the term which gives rise to the rapid dHvA oscillations of M. It shows precisely the expected periodicity – every time $1/B$ changes by an amount $1/F = 2\pi e/\hbar\mathscr{A}_e$, we pass through one cycle of oscillation. The sign of the $\mp\pi/4$ phase factor in θ depends on whether $\mathscr{A}_e$ is a maximum or a minimum area. For a maximum, the rapid change of Δk_z occurs just before $\mathscr{A}_n$ expands through $\mathscr{A}_e$, and the phase factor is $-\pi/4$; for a minimum it occurs just after, and the factor is $+\pi/4$.

It is easy to see the origin of the $|\mathscr{A}''_e|$ term. The oscillations of M will be large if the csa of the FS changes only slowly near the extremal area, so that a large length of the Landau tube empties rapidly, and small if the csa changes rapidly, so that only a short length empties rapidly (problem 7.7).

The $\cos\phi$ term is due to the spin splitting. This adds $\mp\mu_{\text{eff}}B$ to the electron energy, so that if the Fermi energy ε_F is to be the same for both spin-up and spin-down electrons, the highest filled states must have *orbital* energy $\varepsilon_F \pm \mu_{\text{eff}}B$; the FS for the spin-up electrons must correspond to orbital energy $\varepsilon_F + \mu_{\text{eff}}B$, and that for the spin-down electrons to $\varepsilon_F - \mu_{\text{eff}}B$. The extremal orbit areas for the spin-up and spin-down groups are therefore changed from $\mathscr{A}_e$ to $\mathscr{A}_e \pm \mu_{\text{eff}}B(\partial\mathscr{A}/\partial\varepsilon)$, and using (7.7) it is easy to show that this will change $\sin\theta$ to $\sin(\theta \pm \phi)$, where $\phi = \pi\mu_{\text{eff}}m_c^*/\mu_B m = \pi g_{\text{eff}}m_c^*/2m$. The spin-up and spin-down groups of electrons will thus give rise to dHvA signals varying as $\sin(\theta+\phi)$ and $\sin(\theta-\phi)$, and the total signal will be the sum of these two, and will thus vary as $\sin\theta\cos\phi$.

The term $e^{-\alpha T}(=\exp - 2\pi^2 kT/\hbar\omega_c)$ represents the effect of the FS blurring out at finite temperatures. (We have approximated here: the exact form of this term is $\frac{1}{2}\sinh(\alpha T)$, but in practice αT is almost always large enough for $e^{-\alpha T}$ to be a good approximation.) Because of this blurring, the emptying of the nth Landau tube no longer occurs abruptly at $\mathscr{A}_n = \mathscr{A}_e$; instead, it spreads over a range of areas around $\mathscr{A}_e$ corresponding to an energy range $\sim kT$. Clearly if kT is comparable with the energy difference $\hbar\omega_c$ between successive Landau tubes, the effects of successive tubes emptying will begin to overlap, and the resultant fluctuations in the total energy will be very much reduced.

Lastly, the term $e^{-\alpha T_D}$ $(=e^{-\pi/\omega_c\tau})$ represents the collision broadening of the quantized energy levels themselves. If an electron only survives in a quantized orbit for time τ, on average, before being scattered out of it, the energy of the orbit must be broadened, by the uncertainty principle, by an amount $\hbar/\tau$. Just as with thermal broadening of the FS, this collision broadening will diminish the oscillations markedly if $\hbar/\tau$ is comparable with $\hbar\omega_c$. In general, the relaxation time $\tau(\boldsymbol{k})$ may vary around the orbit; if so, the probability of scattering during a time interval δt at point $\boldsymbol{k}$ on the orbit is $\delta t/\tau(\boldsymbol{k})$, and the integral of this around the orbit defines an average scattering rate $1/\tau_{av} = (1/\tau_c)\oint(\mathrm{d}t/\tau(\boldsymbol{k})$. The last term in (7.18) then becomes $\exp - \pi/\omega_c\tau_{av}$.

We can thus learn a great deal from dHvA measurements, and not just about the extremal areas $\mathscr{A}_e$. From (7.18a) we see that the amplitude M_0 of the oscillations is proportional to $TB^{-1/2}e^{-\alpha(T+T_D)}$, so that $\ln(M_0B^{1/2}/T) = C - \alpha(T + T_D) = C - 2\pi^2k(T + T_D)m_c^*/\hbar eB$. Thus a plot of $\ln(M_0/T)$ against T, at fixed value of B, should yield a straight line of slope $-2\pi^2km_c^*/\hbar eB$, from which m_c^* can be determined. (This assumes that T_D does not change as T is varied, i.e. that T is low enough for τ to be dominated by impurity scattering.) A plot of $\ln(M_0B^{1/2})$ against $1/B$, at fixed T, should then yield a straight line of slope $-2\pi^2k(T + T_D)m_c^*/\hbar e$, from which (knowing m_c^*) $T + T_D$ can be determined, and hence T_D and τ_{av} $(= \hbar/2\pi kT_D)$.

If the FS has several different extremal areas for a given field direction, so that the plot of M against $1/B$ is the sum of several signals of different periods, we should be able to distinguish not only the periods of the individual components, from which we can find the extremal areas, but also their individual amplitudes M_0, from which we can then find m_c^* and τ_{av} for each extremal orbit. (In principle, if we can measure the absolute magnitude of M_0 accurately and if we know the value of the spin factor $\cos\phi$, we can also deduce the parameter $\mathscr{A}''_e$, but this has seldom been attempted, both because it is much easier to measure *relative* values of M_0 than absolute values, and because it is not easy to determine the value of $\cos\phi$.)

If we know $\mathscr{A}_e$, m_c^* and τ_{av} for each extremal orbit for a variety of different field directions $\boldsymbol{B}$, we have a vast amount of information available. From the variation of $\mathscr{A}_e$ with field direction, it is in principle possible to deduce the shape of the FS directly, at least in simple cases, by using a rather complicated mathematical 'inversion theorem', but more often the data are compared with the predictions of an approximate band-structure calculation, and used to refine it; by this blend of theory and experiment we can arrive at a very accurate picture of the FS.

Knowing the FS shape, the variation of m_c^* with field direction can be used to deduce, from (7.7), the shape of an adjacent energy surface, and hence to find $\boldsymbol{v}(\boldsymbol{k})$ and its variation over the FS; and finally, knowing $\boldsymbol{v}(\boldsymbol{k})$, the variation of τ_{av} with field direction can be used to find $\tau(\boldsymbol{k})$ and its variation over the FS. However, it is not a simple matter to 'invert' the experimental data on m_c^* and τ_{av} to find the variation of $\boldsymbol{v}(\boldsymbol{k})$ and $\tau(\boldsymbol{k})$ over the FS, and not a great deal of work has been done on this.

Because of the $e^{-\alpha T}$ term, the dHvA effect can normally be observed only at low temperatures, and because of the $e^{-\alpha T_D}$ term, only in reasonably pure samples (problem 7.8), though the effect will remain visible up to higher temperatures if m_c^* is small. The demands on sample purity are in fact not all that stringent, and there has been a good deal of work on the dHvA effect in dilute alloys, to study the effect of alloying on the shape of the FS.

The dHvA effect cannot normally be observed in semiconductors, because at temperatures high enough for the number of carriers (electrons or positive holes) to be appreciable, the term $e^{-\alpha T}$ will make the effect unobservably small. However, some heavily-doped semiconductors retain an appreciable carrier density even at low temperatures, and quantum effects can then be observed. It turns out that in these materials the effects can be much more easily studied as oscillations in the electrical resistivity (the Schubnikov–de Haas effect) than as dHvA oscillations in the susceptibility, but the physical origin is still the same – the quantization of the electron orbits in a field $\boldsymbol{B}$.

8

Collisions

8.1 SCATTERING BY STATIC DEFECTS

In Chapter 7 we neglected the effect of collisions, except to note that in the dHvA effect they would broaden the energy levels and hence diminish the signal. But in determining the transport properties of metals and semiconductors – for example their electrical and thermal conductivities σ and κ – collisions are all-important; if there were no collisions, the relaxation time τ would be infinite, and so therefore would σ and κ.

As we saw in section 7.1, there would indeed be no collisions in a perfect crystal at $T = 0$. In a real crystal at finite temperature, there will be imperfections of two kinds – static, temperature-independent defects, and thermal vibrations of the ions about their lattice sites. The static defects may include dislocations and grain boundaries as well as various kinds of point defect, but we shall consider here only point defects; dislocations and grain boundaries are more difficult to deal with, and they usually contribute only a small amount to the total scattering. The simplest point defects are vacancies, where an atom is missing from a lattice site, and substitutional impurities, where a lattice site is occupied by an impurity atom instead of one of the atoms of the pure crystal. Either of these represents a localized departure from the regular periodic potential of the crystal and will consequently scatter electrons. Because the defect is firmly embedded in the crystal and effectively has infinite mass, the scattering will be elastic (as it will be for other static defects too); that is, an incident electron in state $\boldsymbol{k}$ can be scattered into state $\boldsymbol{k}'$ only if $\varepsilon(\boldsymbol{k}') = \varepsilon(\boldsymbol{k})$.

The scattering of an incident particle by a localized scattering centre is a familiar problem in nuclear and elementary-particle physics, and you may have met it in that context. Here we shall just quote the results of the analysis. If the incident particle is represented by a free-electron plane wave $\psi_k = e^{i\boldsymbol{k}\cdot\boldsymbol{r}}$, and if the scattering centre produces a perturbing potential $V_s(\boldsymbol{r})$ which is not too strong, so that the 'Born approximation' can be used, then the probability per second that the incident particle will be scattered into the state $\psi_{k'} = e^{i\boldsymbol{k}'\cdot\boldsymbol{r}}$ will be given by 'Fermi's golden rule':

$$Q_{\boldsymbol{k}\boldsymbol{k}'} = \frac{2\pi}{\hbar}|V_{\boldsymbol{k}\boldsymbol{k}'}|^2\delta(\varepsilon_{\boldsymbol{k}'} - \varepsilon_{\boldsymbol{k}}) \tag{8.1}$$

where the matrix element

$$V_{\boldsymbol{k}\boldsymbol{k}'} = \int \psi_{\boldsymbol{k}}(\boldsymbol{r})V_s(\boldsymbol{r})\psi^*_{\boldsymbol{k}'}(\boldsymbol{r})\,\mathrm{d}^3\boldsymbol{r} \tag{8.2}$$

$$= \int e^{i(\boldsymbol{k}-\boldsymbol{k}')\cdot\boldsymbol{r}}V_s(\boldsymbol{r})\,\mathrm{d}^3\boldsymbol{r} \tag{8.3}$$

is just the Fourier transform of $V_s(\boldsymbol{r})$, and $\delta(\varepsilon_{k'} - \varepsilon_k)$ is a delta-function such that $\delta(\varepsilon_{k'} - \varepsilon_k) = 0$ unless $\varepsilon_{k'} = \varepsilon_k$, and $\int \delta(\varepsilon_{k'} - \varepsilon_k)\,\mathrm{d}\varepsilon_{k'} = 1$. (From now on, we shall usually write $V_{kk'}$, ε_k, etc. in place of the more cumbersome $V(\boldsymbol{k},\boldsymbol{k}')$, $\varepsilon(\boldsymbol{k})$, etc.)

The number of states available in volume $\delta^3\boldsymbol{k}'$, per unit volume of real space, is $\delta^3\boldsymbol{k}'/4\pi^3$, from (1.23). But this includes both spin-up and spin-down states; scattering by impurities or by thermal vibrations will normally leave the spin unchanged, so the number of states that $\boldsymbol{k}$ can scatter into will be $\delta^3\boldsymbol{k}'/8\pi^3$. If we assume that there are n_i impurity scatterers per unit volume, acting independently, then the total probability per second of scattering into the group of states $\delta^3\boldsymbol{k}'$, $P_{kk'}\delta^3\boldsymbol{k}'$, is given by

$$P_{kk'}\delta^3\boldsymbol{k}' = n_i Q_{kk'}\delta^3\boldsymbol{k}'/8\pi^3 = (n_i/4\pi^2\hbar)|V_{kk'}|^2\delta(\varepsilon_{k'} - \varepsilon_k)\delta^3\boldsymbol{k}' \tag{8.4}$$

(This result is in fact independent of the size of the system; if the real-space volume is V_r, the wave-functions become $\psi_k = V_r^{-1/2}e^{i\boldsymbol{k}\cdot\boldsymbol{r}}$, etc. when normalized, so that $|V_{kk'}|^2 \propto V_r^{-2}$; but for a given impurity concentration the total number of scatterers becomes $n_i V_r$, and the number of states in $\delta^3\boldsymbol{k}'$ becomes $V_r\delta^3\boldsymbol{k}'/8\pi^3$, so that the V_r factors just cancel.) If we write $\delta^3\boldsymbol{k}' = \delta k'_\perp\,\delta S_{k'} = \delta S_{k'}\delta\varepsilon_{k'}/\hbar v(\boldsymbol{k}')$, as in deriving (7.2), and use $\int \delta(\varepsilon_{k'} - \varepsilon_k)\,\mathrm{d}\varepsilon_{k'} = 1$, (8.4) integrates to give

$$\int P_{kk'}\,\mathrm{d}^3\boldsymbol{k}' = (n_i/4\pi^2\hbar^2)\int |V_{kk'}|^2\,\mathrm{d}S_{k'}/v(\boldsymbol{k}') \tag{8.5}$$

where the integration is over the constant-energy surface $\varepsilon = \varepsilon(\boldsymbol{k})$.

In principle, (8.4) or (8.5) should be applicable to the scattering of Bloch electrons as well as free electrons: the wave-functions in (8.2) will then be Bloch functions instead of plane waves, and $V_s(\boldsymbol{r})$ will be the *difference* between the potential in a crystal containing a single impurity atom (or vacancy) and in a perfect crystal. Clearly this difference will vanish except close to the impurity. In practice, $V_s(\boldsymbol{r})$ may well be too strong for the Born approximation to be valid, but for simple metals we can instead take $V_s(\boldsymbol{r})$ to be the (much weaker) difference in *pseudopotentials* between the crystal with and without the impurity atom. The functions ψ_k, $\psi_{k'}$ in (8.2) will then be the pseudo-wave-functions (5.5), and these can often be well approximated by a single plane wave, so that (8.2) still reduces to the simpler form (8.3).

(For transition metal impurities, $V_s(\boldsymbol{r})$ may be strong enough to form a 'virtual bound state' around the impurity, which scatters so strongly that the Born approximation is no longer useful. One then has to use the more powerful methods of phase shift analysis to tackle the problem, but this would take us too far afield.)

To get an idea of the form of $V_{kk'}$, consider a metal in which the FS does not touch the BZ boundary, and is therefore a sphere of radius k_F in the NFE approximation. For scattering between states on the FS, we therefore have $|\boldsymbol{k}| = |\boldsymbol{k}'| = k_F$, and if $V_s(\boldsymbol{r})$ is spherically symmetric, (8.3) reduces to

$$V_{kk'} = (4\pi/k_s)\int_0^\infty r\,V_s(r)\sin(k_s r)\,\mathrm{d}r \tag{8.6}$$

where $k_s = |\boldsymbol{k} - \boldsymbol{k}'|$; if θ is the angle between $\boldsymbol{k}$ and $\boldsymbol{k}'$, $k_s = 2k_F \sin\frac{1}{2}\theta$. Consider two alternative forms of $V_s(r)$:

(a) $$V_s(r) = (\Delta Z e^2/4\pi\varepsilon_0 r)\,\mathrm{e}^{-r/r_1}; \tag{8.7}$$

(b) $$V_s(r) = V_0 \text{ for } r < r_0, \quad V_s(r) = 0 \text{ for } r \geqslant r_0 \tag{8.8}$$

The first of these might represent, very roughly, scattering by an impurity atom whose valency differs by ΔZ from that of a 'pure' atom. (We can also apply it to scattering by a vacancy, by regarding the vacancy as effectively an impurity atom of zero valency.) For example, a Pb atom, present as an impurity in Al, will contribute four valence electrons to the conduction band, leaving the Pb ion with a charge of $+4e$, whereas the Al ions have a charge of $+3e$. The additional positive charge on the Pb ion will produce an additional attractive potential around it, which we might expect to look like $-e^2/4\pi\varepsilon_0 r$. But precisely because the potential *is* attractive, there will be a higher density of electrons around this ion, and they will tend to screen it so that the potential drops off more rapidly. In fact it drops off exponentially, as in (8.7), with a 'screening length' r_1. Starting from $\mathrm{div}(\varepsilon_0 \boldsymbol{E}) = \rho$ (where ρ is the charge density), it is not difficult to show that in metals

$$1/r_1^2 = e^2 g(\varepsilon_\mathrm{F})/\varepsilon_0 \tag{8.9}$$

and hence that $\boldsymbol{r}_1$ is about 1 Å (0.1 nm). In semiconductors,

$$1/r_1^2 = e^2 n/\varepsilon_\mathrm{r}\varepsilon_0 kT \tag{8.10}$$

where n is the number of charge carriers per unit volume and ε_r is the dielectric constant of the material, so that r_1 is much larger in semiconductors – typically about 1000 Å.

(Exactly the same kind of screening occurs, of course, when an electrostatic field is applied to the surface of a metal or a semiconductor. The conduction electrons are attracted to the surface, and screen out the applied field so that it falls off as e^{-x/r_1} below the surface.)

Equation (8.7) is only a crude approximation because, among other things, it treats the ions as point charges, which they certainly are not. It fails to predict any scattering at all if the impurity has the same valency as the pure atom, and experimentally the scattering from such 'homovalent' impurities is indeed small (as indicated for example by the residual resistivity), but it is not negligible, because there is still a small difference in the effective pseudopotential between the impurity and the pure atom. Equation (8.8) can be used to represent this, in a rough and ready way.

Inserting (8.7) or (8.8) in (8.6), we find that $V_{\boldsymbol{kk}'}/V_{\boldsymbol{kk}'}(0)$ varies with k_s as shown in Fig. 8.1 (see problem 8.2). Here $V_{\boldsymbol{kk}'}(0)$ is the value of $V_{\boldsymbol{kk}'}$

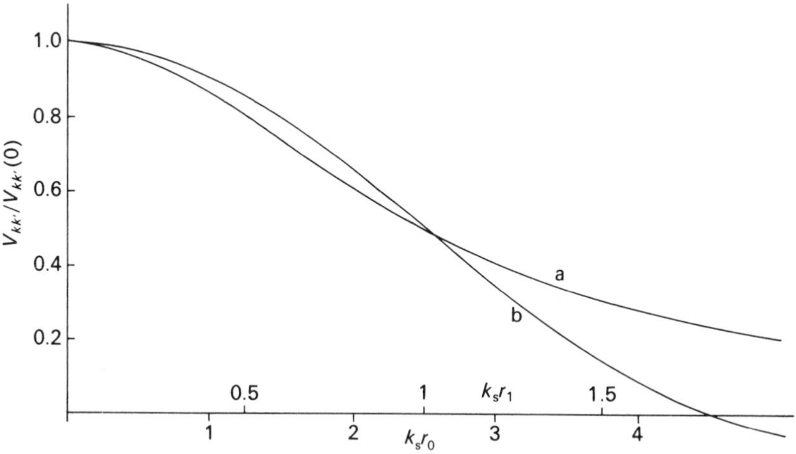

Fig. 8.1 The variation of the matrix element $V_{kk'}$ with (a) $k_s r_1$, for the potential (8.7); (b) $k_s r_0$, for (8.8).

at $k_s = 0$, and is given by

$$\text{(a) } V_{kk'}(0) = \Delta Z e^2 r_1^2/\varepsilon_0; \qquad \text{(b) } V_{kk'}(0) = (4\pi r_0^3/3)V_0 \qquad (8.11)$$

Figure 8.1 shows that $V_{kk'}$ falls off at large values of k_s and hence of scattering angle θ; the electron is more likely to be scattered through

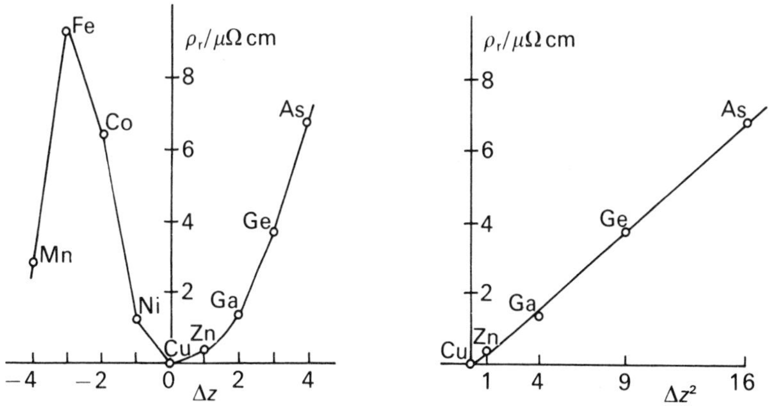

Fig. 8.2 The residual resistivity of Cu containing 1 atomic per cent of different impurities, illustrating Linde's rule. [From Linde (1932), *Ann. der Phys*, **15**, 219.]

small angles than large ones. Since $P_{kk'} \propto |V_{kk'}|^2$, the scattering (if $\Delta Z \neq 0$) is proportional to ΔZ^2, according to (8.11), and this agrees with experiment; the residual resistivity, for a given impurity concentration n_i, does indeed vary as ΔZ^2 for non-transition metal impurities, as shown in Fig. 8.2. This is known as Linde's rule. For transition metal impurities (those of negative ΔZ in Fig. 8.2) the rule fails, because such impurities scatter too strongly for the Born approximation to be useful, as we noted earlier.

8.2 PHONON SCATTERING

At finite temperatures, the thermal vibrations of the ions produce departures from the perfect regularity of the ideal crystal, and hence lead to scattering. The thermal vibrations are described in terms of phonons (quantized sound waves): a phonon of wave-vector $\boldsymbol{q}$ and angular frequency $\omega_{\boldsymbol{q}}$ will have energy $\hbar\omega_{\boldsymbol{q}}$, and in thermal equilibrium the number $n_{\boldsymbol{q}}$ of such phonons will be given by the Bose–Einstein function

$$n_{\boldsymbol{q}} = \frac{1}{\exp(\hbar\omega_{\boldsymbol{q}}/kT) - 1} \tag{8.12}$$

The $\omega(\boldsymbol{q})$ spectrum, like the $\varepsilon(\boldsymbol{k})$ spectrum for electrons, repeats periodically throughout $\boldsymbol{q}$-space (which is the same thing as $\boldsymbol{k}$-space), so that all the physically relevant information is contained in the central BZ. Near the centre of the zone, where $\boldsymbol{q}$ is small, we have $\omega_{\boldsymbol{q}} = v_s q$, where v_s is the speed of sound; further out, $\omega_{\boldsymbol{q}}$ rises more slowly. The highest phonon frequency in the BZ, ω_{max}, can be estimated roughly from the relation $\hbar\omega_{max} \approx k\theta_D$, where θ_D is the Debye temperature deduced from heat capacity measurements. (These two quantities would be equal on the Debye theory, but the Debye theory assumes a very simplified phonon spectrum.) We shall consider here only longitudinal (i.e. longitudinally polarized) phonons; transverse phonons also exist, but they usually scatter electrons much more weakly, and we can neglect them.

When an electron is scattered from $\boldsymbol{k}$ to $\boldsymbol{k}'$ by the thermal vibrations, it either emits or absorbs a phonon, so that the scattering in inelastic; energy conservation requires that

$$\varepsilon_{\boldsymbol{k}'} = \varepsilon_{\boldsymbol{k}} \pm \hbar\omega_{\boldsymbol{q}} \tag{8.13}$$

(+ for absorption, − for emission). The scattering must also satisfy what is essentially a Bragg diffraction condition (though it is sometimes called a momentum conservation condition):

$$\boldsymbol{k}' = \boldsymbol{k} \pm \boldsymbol{q} + \boldsymbol{K} \tag{8.14}$$

(+ for absorption, − for emission), where $\boldsymbol{K}$ is a reciprocal lattice vector. This vector allows for the inherent non-uniqueness of $\boldsymbol{k}$ and $\boldsymbol{k}'$, as discussed in Chapter 4 (cf. (4.7), (4.8)); this non-uniqueness is shared by $\boldsymbol{q}$, as shown by the periodicity of $\omega(\boldsymbol{q})$. But even if we resolve this non-uniqueness by choosing $\boldsymbol{k}$, $\boldsymbol{k}'$ and $\boldsymbol{q}$ to lie within the central BZ (as we normally do), we still need to include $\boldsymbol{K}$ in (8.14), because the vector $\boldsymbol{k}' - \boldsymbol{k}$ can still lie outside the central BZ. If $\boldsymbol{k}' - \boldsymbol{k}$ lies inside the BZ, (8.14) can be satisfied with $\boldsymbol{K} = 0$, and we have what is called a Normal process or N-process; if not, we have $\boldsymbol{K} \neq 0$, and what is called an 'Umklapp' or U-process.

The expression for phonon scattering is very similar in form to the expression (8.4) for impurity scattering, except that we now have two separate terms, one for phonon-emission processes and one for phonon-absorption processes:

$$P_{\boldsymbol{kk}'}\delta^3\boldsymbol{k}' = [(n_{\boldsymbol{q}} + 1)Q^-_{\boldsymbol{kk}'} + n_{\boldsymbol{q}}Q^+_{\boldsymbol{kk}'}]\delta^3\boldsymbol{k}'/8\pi^3 \tag{8.15}$$

where

$$Q^{\pm}_{\boldsymbol{kk}'} = \frac{2\pi}{\hbar}|V_{\boldsymbol{kk}'}|^2\delta[\varepsilon_{\boldsymbol{k}'} - (\varepsilon_{\boldsymbol{k}} \pm \hbar\omega_{\boldsymbol{q}})] \tag{8.16}$$

The factor $n_{\boldsymbol{q}} + 1$ rather than $n_{\boldsymbol{q}}$ appears in the phonon-emission term for exactly the same reason that it appears in the theory of photon emission by an excited atom (in which case $n_{\boldsymbol{q}}$ is the number of *photons* $\boldsymbol{q}$): the $n_{\boldsymbol{q}}$ part represents stimulated emission, and the extra $+1$ represents spontaneous emission. As it happens, the $+1$ is not very important in practice, and we can often neglect it. (We might expect it to be important when $n_{\boldsymbol{q}} \gtrsim 1$, i.e. when $\hbar\omega_{\boldsymbol{q}} > kT$, from (8.12), but then the exclusion principle will largely inhibit phonon-emission processes anyway, because states $\boldsymbol{k}'$ of energy $\varepsilon_{\boldsymbol{k}} - \hbar\omega_{\boldsymbol{q}}$ will already be full. For a more detailed discussion of this approximation, see Appendix B.)

We still have to estimate $|V_{\boldsymbol{kk}'}|^2$, and we can do this by thinking of the phonon semi-classically, as a wave travelling through the crystal and displacing the ion at $\boldsymbol{r}$ by an amount

$$\boldsymbol{u} = \boldsymbol{u}_0 \exp i(\boldsymbol{q}\cdot\boldsymbol{r} - \omega_{\boldsymbol{q}}t)$$

The ion (of mass M) thus has a kinetic energy which oscillates between zero and $\frac{1}{2}M\omega_q^2 u_0^2$. Its potential energy correspondingly oscillates between $\frac{1}{2}M\omega_q^2 u_0^2$ and zero, so that its total energy, due to the wave, is $\frac{1}{2}M\omega_q^2 u_0^2$. If we consider a crystal of unit volume, containing N ions, we can thus find the amplitude of vibration u_0 for a single phonon by writing

$$\hbar\omega_q = \tfrac{1}{2}NM\omega_q^2 u_0^2 \tag{8.17}$$

(For n_q phonons, u_0^2 will be n_q times as large, but that is already taken into account in (8.15).) Now if every ion in the crystal were displaced by the same amount $\boldsymbol{u}$, the crystal would remain perfect; what produces the scattering is the deformation due to the *variation* of $\boldsymbol{u}$ with position, and particularly the compressional deformation, div $\boldsymbol{u} = i\,\boldsymbol{q}\cdot\boldsymbol{u}$. (For transverse phonons, $\boldsymbol{q}\cdot\boldsymbol{u} = 0$, which is why we can to good approximation neglect them.) We thus expect to find $|V_{\boldsymbol{kk}'}|^2 \propto q^2 u_0^2$, with u_0^2 given by (8.17). Detailed theory confirms this, and leads to the result

$$|V_{\boldsymbol{kk}'}|^2 = V_0^2 \hbar q^2/2NM\omega_q \tag{8.18}$$

where $V_0 \approx 2\varepsilon_F/3$ for an NFE metal and $\boldsymbol{k}, \boldsymbol{k}'$ and $\boldsymbol{q}$ are related by (8.14) It follows that for phonon scattering, as for impurity scattering, $P_{\boldsymbol{kk}'}$ is independent of the volume V_r of the crystal; in (8.15), as in (8.4), $\delta^3\boldsymbol{k}'/8\pi^3$ must be replaced by $V_r\delta^3\boldsymbol{k}'/8\pi^3$ for a crystal of volume V_r and, in (8.17) and (8.18), N must be replaced by NV_r, so that the V_r factors again cancel. When proper account is taken of the V_r factors, both (8.4) and (8.15) have the dimensions of time^{-1}, as expected (problem 8.3).

We see that for phonon scattering, as for impurity scattering, $|V_{\boldsymbol{kk}'}|^2$ and $P_{\boldsymbol{kk}'}$ will vary with $\boldsymbol{k}' - \boldsymbol{k}$ and hence with the scattering angle. For N-processes (i.e. for $\boldsymbol{K} = 0$ in (8.14)), we have $\boldsymbol{k}' - \boldsymbol{k} = \pm\boldsymbol{q}$, and for small q we can write $\omega_q = v_s q$, so that $|V_{\boldsymbol{kk}'}|^2 \propto q$; unlike impurity scattering, phonon scattering vanishes for $|\boldsymbol{k}' - \boldsymbol{k}| \to 0$. As $\boldsymbol{k}' - \boldsymbol{k}$ grows larger, $\boldsymbol{q}$ will likewise grow larger, until it reaches the BZ boundary and passes through it into the next BZ. As it does so, there is no abrupt change in the physics of the scattering process, because in the periodic zone scheme $\omega(\boldsymbol{q})$, like $\varepsilon(\boldsymbol{k})$, repeats itself periodically in $\boldsymbol{q}$-space, and continues to vary smoothly with $\boldsymbol{q}$ as we pass from one BZ to the next. Indeed, we could continue to write $\boldsymbol{k}' - \boldsymbol{k} = \boldsymbol{q}$, if we were willing to let $\boldsymbol{q}$ extend outside the central BZ. But if we want to confine

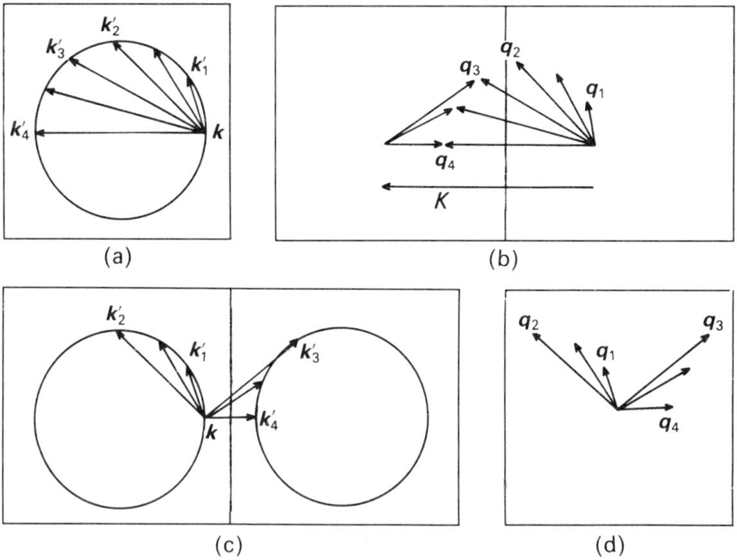

Fig. 8.3 (a) The vectors $\boldsymbol{k}' - \boldsymbol{k}$ involved in scattering from $\boldsymbol{k}$ to various points $\boldsymbol{k}'$. (b) The corresponding vectors $\boldsymbol{q}, \boldsymbol{q}_3$ and $\boldsymbol{q}_4$ extend outside the BZ, unless we subtract $\boldsymbol{K}$. (c) Alternatively, if states $\boldsymbol{k}'_3$ and $\boldsymbol{k}'_4$ are replotted in the next BZ, displaced by $\boldsymbol{K}$, $\boldsymbol{q}_3$ and $\boldsymbol{q}_4$ stay in the central BZ, as shown in (d).

$\boldsymbol{q}$ (along with $\boldsymbol{k}$ and $\boldsymbol{k}'$) to the central BZ, we must replace $\boldsymbol{q} = \boldsymbol{k}' - \boldsymbol{k}$ by the completely equivalent vector $\boldsymbol{q} = \boldsymbol{k}' - \boldsymbol{k} - \boldsymbol{K}$ as soon as $\boldsymbol{q}$ crosses the BZ boundary, so that N-processes are replaced by U-processes at this point, as shown in Fig. 8.3b. Clearly, there is no sudden change in the physics at this point; merely in our way of describing it mathematically.

The importance of U-processes is that as $\boldsymbol{k}' - \boldsymbol{k}$ increases still further, $\boldsymbol{q}$ now decreases (Fig. 8.3b), and so therefore does $\omega_{\boldsymbol{q}}$. Now from (8.12), the number of phonons $n_{\boldsymbol{q}}$ increases as $\omega_{\boldsymbol{q}}$ decreases, and since $P_{\boldsymbol{kk}'}$ depends on $n_{\boldsymbol{q}}$, this means that $P_{\boldsymbol{kk}'}$ may be larger for large values of $\boldsymbol{k}' - \boldsymbol{k}$ than for intermediate values. (Hence the German term 'Umklapp': a small $\boldsymbol{q}$ can 'turn over' the direction of $\boldsymbol{k}$.)

An alternative way of showing N-processes and U-processes is to make use of the periodic zone scheme for $\varepsilon(\boldsymbol{k})$, as shown in Fig. 8.3c. The vectors $\boldsymbol{q}_3$ and $\boldsymbol{q}_4$ now connect $\boldsymbol{k}$ in the central BZ to $\boldsymbol{k}'_3$ and $\boldsymbol{k}'_4$ in the adjacent BZ.

So far we have neglected the energy conservation condition (8.13). Because of this, the state $\boldsymbol{k}'$ will lie a little way off the constant-energy surface $\varepsilon = \varepsilon_k$, though the displacement is too small to show in Fig. 8.3: the energy $\hbar\omega_q$ will typically be ~ 0.05 eV at most (remembering that $\hbar\omega_{\max} \sim k\theta_D$), very small on the scale of electron energies ε_k, which typically vary by 5–10 eV between the bottom and the top of a band. For metals, indeed, we can often neglect the fact that the collisions are inelastic, and replace $\delta[\varepsilon_{k'} - (\varepsilon_k \pm \hbar\omega_q)]$ in (8.16) by $\delta(\varepsilon_{k'} - \varepsilon_k)$. If we do this, and if we also approximate by writing n_q in place of $n_q + 1$, the two terms in (8.15) become equal, and we can rewrite that equation in a similar form to (8.5):

$$\int P_{kk'}\,\mathrm{d}^3\boldsymbol{k}' = (2n_q/4\pi^2\hbar^2)\int \mathrm{d}S_{k'}|V_{kk'}|^2/v_{k'} \tag{8.19}$$

In any real crystal, both static-defect scattering and phonon scattering will be present simultaneously, and the total scattering rate for $\boldsymbol{k}$ into $\delta^3\boldsymbol{k}'$ will be the sum of the two separate contributions, (8.4) and (8.15) or (on integration) (8.5) and (8.19).

8.3 RELAXATION TIMES AND MEAN FREE PATHS

If $P_{kk'}\,\delta^3\boldsymbol{k}'$ is the probability per second of scattering from state $\boldsymbol{k}$ to states in the volume element $\delta^3\boldsymbol{k}'$, then

$$1/\tau_k^0 = \int P_{kk'}\,\mathrm{d}^3\boldsymbol{k}' \tag{8.20}$$

gives the total probability per second of scattering out of state $\boldsymbol{k}$ into all other states. To see the significance of τ_k^0, suppose that at time $t = 0$ there are say $N(0)$ electrons in a volume element $\delta^3\boldsymbol{k}$ around $\boldsymbol{k}$, and that $N(t)$ of these survive without scattering until time t. Between t and $t + \delta t$, each of these survivors has a probability $\delta t/\tau_k^0$ of being scattered, so that the change in $N(t)$ in time δt is

$$\delta N(t) = -N(t)\,\delta t/\tau_k^0$$

and hence

$$N(t) = N(0)\,\mathrm{e}^{-t/\tau_k^0} \tag{8.21}$$

It follows that of the original $N(0)$ electrons, a fraction $p(t)\delta t = \mathrm{e}^{-t/\tau_k^0}\,\delta t/\tau_k^0$ we survive until time t and then collide in the interval δt, so that the average survival time is

$$\bar{t} = \int_0^\infty t\, p(t)\,\mathrm{d}t = \tau_k^0 \int_0^\infty x\,\mathrm{e}^{-x}\,\mathrm{d}x = \tau_k^0 \tag{8.22}$$

(putting $x = t/\tau_k^0$). Now this result is independent of the history of the electrons before $t = 0$; in particular, an electron which was scattered into state $\boldsymbol{k}$ immediately before $t = 0$ will on average survive for a time τ_k^0 before being scattered out. τ_k^0 is therefore just the average time between collisions – in other words, it is just the relaxation time τ, as defined in Chapter 1, for electrons in state $\boldsymbol{k}$.

We remarked in Chapter 1 that if at time $t = 0$ we pick an electron at random (instead of one which has just collided), it is likely on average to be half-way between collisions, and therefore to have already travelled for a time τ since its last collision. This led to the apparent paradox that for an electron chosen at random, the mean time between collisions is 2τ rather than τ, and we explained this by saying that an electron chosen at random was more likely to be in the midst of a long free path than a short one. We can now confirm this by working out the appropriate weighted average, and it does indeed turn out to be 2τ (problem 8.4).

In Chapter 1, we then argued that in an applied field $\boldsymbol{E}$, an electron will on average acquire a drift velocity $-e\boldsymbol{E}\tau/m$, since it will on average have been travelling for a time τ since its last collision. But that argument implicitly assumed that after each collision the electron starts out completely afresh, with no memory of any drift velocity it might have acquired previously. This will not usually be true. In particular, collisions which only change $\boldsymbol{k}$ and $\boldsymbol{v}$ by a small amount will do little to randomize the electron's motion or to limit the drift velocity it acquires. Only if it is equally likely to emerge in any direction from each collision – in other words, only if $P_{kk'}$ is the same for all directions of $\boldsymbol{k}'$ – will each collision be truly randomizing, and only then will τ_k^0 adequately describe the collision process. If we want to find the electrical conductivity σ, for example, we shall see in Chapter 9 that what we need to know is not τ_k^0 but the vector mean free path $\boldsymbol{L}_k$ – the average *distance* an electron travels, starting in state $\boldsymbol{k}$, before its motion becomes completely randomized. If every collision is truly randomizing, we shall have $\boldsymbol{L}_k = \boldsymbol{v}_k\tau_k^0$, but if each collision only changes the direction of $\boldsymbol{v}$ slightly, $\boldsymbol{L}_k$ may be much larger.

We can derive an expression for $\boldsymbol{L}_k$ as follows. An electron will on average stay in state $\boldsymbol{k}$ for a time τ_k^0 before colliding, and in that time it will travel a distance $\boldsymbol{v}_k\tau_k^0$. On collision, it will be scattered into some other state $\boldsymbol{k}'$, and in *that* state its vector mean free path will be $\boldsymbol{L}_{k'}$. Now the probability that it will be scattered into any particular state $\boldsymbol{k}'$ (or into volume element $\delta^3\boldsymbol{k}'$ around $\boldsymbol{k}'$), out of all the states $\boldsymbol{k}'$ available, will be

$$P_{kk'}\delta^3\boldsymbol{k}' \Big/ \int P_{kk'}\mathrm{d}^3\boldsymbol{k}' = \tau_k^0 P_{kk'}\delta^3\boldsymbol{k}'$$

so that the average distance it travels after being scattered out of state $\boldsymbol{k}$ is given by averaging $\boldsymbol{L}_{k'}$ over all states $\boldsymbol{k}'$, with this weighting factor. Adding the result to $\boldsymbol{v}_k\tau_k^0$, we find

$$\boldsymbol{L}_k = \boldsymbol{v}_k\tau_k^0 + \tau_k^0\int \mathrm{d}^3\boldsymbol{k}' P_{kk'}\boldsymbol{L}_{k'} \tag{8.23}$$

This integral equation for $\boldsymbol{L}_k$, though easy to derive, is not so easy to solve, but we can at least see that it behaves sensibly. Because the energy surface $\varepsilon(\boldsymbol{k})$ always has a centre of symmetry (i.e. $\varepsilon_{-k} = \varepsilon_k$), it follows that $\boldsymbol{v}_{-k} = -\boldsymbol{v}_k$, and in fact that $\boldsymbol{L}_{-k} = -\boldsymbol{L}_k$. If each collision is truly randomizing, so that $P_{kk'} = P_{k,-k'}$, the integral in (8.23) will therefore vanish, because the electron is equally likely to be scattered into states $\boldsymbol{k}'$ and $-\boldsymbol{k}'$, with equal and opposite values of $\boldsymbol{L}_{k'}$. We then have $\boldsymbol{L}_k = \boldsymbol{v}_k\tau_k^0$, as expected. On the other hand, if scattering is predominantly through small angles, the integral term will be only slightly smaller than $\boldsymbol{L}_k$ itself, and it follows that $\boldsymbol{L}_k$ can become much larger than $\boldsymbol{v}_k\tau_k^0$.

$\boldsymbol{L}_k$ may not be exactly parallel to $\boldsymbol{v}_k$, if $P_{kk'}$ tends to scatter the electrons more to one side than the other, but if for simplicity we assume that they *are* parallel, we can write $\boldsymbol{L}_k = \boldsymbol{v}_k\tau_k^{\mathrm{e}}$, where τ_k^{e} is an 'effective' relaxation time. If we take the scalar product $\boldsymbol{v}_k\cdot\boldsymbol{L}_k$, (8.23) then gives

$$v_k^2\tau_k^{\mathrm{e}} = \tau_k^0\left(v_k^2 + \int \mathrm{d}^3\boldsymbol{k}' P_{kk'}\boldsymbol{v}_k\cdot\boldsymbol{v}_{k'}\tau_{k'}^{\mathrm{e}}\right)$$

or if we divide through by $v_k^2\tau_k^0$ and use (8.20),

$$\tau_k^{\mathrm{e}}\int P_{kk'}\mathrm{d}^3\boldsymbol{k}' = 1 + \int \mathrm{d}^3\boldsymbol{k}' P_{kk'}(v_{k'}/v_k)\cos\theta\,\tau_{k'}^{\mathrm{e}} \tag{8.24}$$

where θ is the angle between $\boldsymbol{v}_k$ and $\boldsymbol{v}_{k'}$.

If we further assume that $\tau_{k'}^{\mathrm{e}} = \tau_{k}^{\mathrm{e}}$ and that $v_{k'} = v_{k}$ (which should be reasonably good approximations if the scattering is mainly through small angles), (8.24) reduces to

$$1/\tau_{k}^{\mathrm{e}} = \int \mathrm{d}^3\boldsymbol{k}' P_{\boldsymbol{k}\boldsymbol{k}'} (1 - \cos\theta). \tag{8.25}$$

If $P_{\boldsymbol{k}\boldsymbol{k}'} = P_{\boldsymbol{k},-\boldsymbol{k}'}$, so that each collision is fully randomizing, the $\cos\theta$ term here will integrate to zero, since θ differs by π for states $\boldsymbol{k}'$ and $-\boldsymbol{k}'$. We then have $\tau_{k}^{\mathrm{e}} = \tau_{k}^{0}$, as expected. But if scattering is mainly through small angles, for which $\cos\theta \approx 1$, τ_{k}^{e} may be much greater than τ_{k}^{0}; the $(1 - \cos\theta)$ factor allows for the ineffectiveness of small-angle collisions in randomizing the electron's velocity.

On the other hand, even small-angle collisions, if they are inelastic, may be quite effective in randomizing the electron's *energy*. As we shall see in Chapter 10, this means that the thermal conductivity is determined by a different effective relaxation time τ_{k}^{T}, which may be much less than τ_{k}^{e}, and may indeed be closer to τ_{k}^{0}.

9

Electrical conductivity of metals

9.1 THE BASIC EXPRESSION FOR σ

Now that we have looked at the effect of electric fields and of collisions separately, we can look at the effect of the two together, and see how they determine the conductivity and other transport properties. We consider metals in this and the following three chapters, and then turn to semiconductors.

In thermal equilibrium, the probability of state $\boldsymbol{k}$ being occupied is given by the Fermi–Dirac function $f_0(\varepsilon_{\boldsymbol{k}})$ of equation (1.11), which we can write as $f_0(\boldsymbol{k})$ or $f_{0,\boldsymbol{k}}$ for short. The effect of an applied electric field, or of a temperature gradient, is to move the system away from equilibrium, so that the probability of state $\boldsymbol{k}$ being occupied is changed to $f(\boldsymbol{k})$ say. As in Chapter 7, we shall think of the electrons as localized wave-packets, because we want to be able to discuss situations – for example, a metal in a temperature gradient, or in a non-uniform electric field – which are not spatially uniform. We can then write $f = f(\boldsymbol{k}, \boldsymbol{r})$, so that (from (1.23)) the number of electrons in volume element $\delta^3\boldsymbol{k}$ of $\boldsymbol{k}$-space and $\delta^3\boldsymbol{r}$ of real space, at point $\boldsymbol{k}, \boldsymbol{r}$, will be

$$\delta n(\boldsymbol{k}, \boldsymbol{r}) = f(\boldsymbol{k}, \boldsymbol{r})\delta^3\boldsymbol{k}\delta^3\boldsymbol{r}/4\pi^3 \tag{9.1}$$

The number in $\delta^3\boldsymbol{k}$, per *unit volume* of real space, will be $f(\boldsymbol{k}, \boldsymbol{r})\delta^3\boldsymbol{k}/4\pi^3$, so that we can find the current density $\boldsymbol{J}(\boldsymbol{r})$, for example, by writing

$$\boldsymbol{J}(\boldsymbol{r}) = -(e/4\pi^3)\int \boldsymbol{v}(\boldsymbol{k}) f(\boldsymbol{k}, \boldsymbol{r})\, \mathrm{d}^3\boldsymbol{k} \tag{9.2}$$

which is a generalized version of (2.39). Clearly $\boldsymbol{J}(\boldsymbol{r})$ must vanish in thermal equilibrium, when $f = f_0$, and so it does, because

$$f_0(\boldsymbol{k}) = f_0(-\boldsymbol{k}) \quad \text{and} \quad \boldsymbol{v}(\boldsymbol{k}) = -\boldsymbol{v}(-\boldsymbol{k}).$$

We can thus replace f by $f_1 = f - f_0$ in (9.2), since the contribution from f_0 vanishes.

In any transport problem, then, our task is to calculate $f(\boldsymbol{k}, \boldsymbol{r})$; we can then find $\boldsymbol{J}(\boldsymbol{r})$ from (9.2), with a similar expression ((10.1) below) for the heat current density $\boldsymbol{Q}(\boldsymbol{r})$. One way of calculating $f(\boldsymbol{k}, \boldsymbol{r})$ is to use the *Boltzmann equation*, and we describe this approach briefly in Appendix B. We shall use here an alternative 'path-integral' approach, which yields the same results rather more simply. Whichever approach one uses, it is difficult to treat the scattering process exactly except in the simplest problems, and one very often uses the 'relaxation time approximation' instead. In this approximation, the actual scattering process is replaced by one in which every scattering event is assumed to be fully randomizing, though the scattering rate is not necessarily assumed to be the rate $1/\tau_k^0$ of (8.20). Usually, indeed, the scattering rate $1/\tau_k$ or the equivalent relaxation time τ_k is treated as an adjustable parameter, to be chosen to fit the experiments. It would not be very satisfactory if this was the best we could do in calculating the basic quantities σ and κ, and here indeed we can do better, and treat the scattering process more or less exactly. To begin with, though, let us calculate σ using the relaxation time approximation.

In Chapter 2 we calculated σ by finding the drift velocity $\boldsymbol{v}_d$ produced by a uniform field $\boldsymbol{E}$. For Bloch electrons, it is easier to think in terms of $\boldsymbol{k}$ rather than $\boldsymbol{v}$, and in section 7.1 we saw that a uniform field $\boldsymbol{E}$, acting for a time τ, would displace an electron in $\boldsymbol{k}$-space by the (extremely small) amount $\boldsymbol{k}_d = -e\boldsymbol{E}\tau/\hbar$. Thus an electron which has (on average) been travelling for time τ since its last collision must have started out from that collision in state $\boldsymbol{k} - \boldsymbol{k}_d$, if it is now in state $\boldsymbol{k}$. The essential simplifying assumption made in the relaxation time approximation is that every collision is truly randomizing, so that the electron starts out from each collision with no memory of its previous history, and thus with the distribution function f_0. We can therefore write

$$f(\boldsymbol{k}) = f_0(\boldsymbol{k} - \boldsymbol{k}_d) \tag{9.3}$$

and because $\boldsymbol{k}_d$ is so small this can be written

$$\begin{aligned} f(\boldsymbol{k}) &= f_0(\boldsymbol{k}) - \boldsymbol{k}_d \cdot \frac{\partial f_0}{\partial \boldsymbol{k}} \\ &= f_0(\boldsymbol{k}) - \boldsymbol{k}_d \cdot \frac{\partial \varepsilon}{\partial \boldsymbol{k}} \frac{\partial f_0}{\partial \varepsilon} \end{aligned}$$

since f_0 depends on $\boldsymbol{k}$ only through ε_k. In other words,

$$f(\boldsymbol{k}) = f_0(\boldsymbol{k}) - \hbar \boldsymbol{k}_{\mathrm{d}} \cdot \boldsymbol{v}_k \frac{\partial f_0}{\partial \varepsilon}$$

$$= f_0(\boldsymbol{k}) + e\tau_k \boldsymbol{E} \cdot \boldsymbol{v}_k \frac{\partial f_0}{\partial \varepsilon}$$

$$= f_0(\boldsymbol{k}) + e\boldsymbol{E} \cdot \boldsymbol{L}_k \frac{\partial f_0}{\partial \varepsilon} \tag{9.4}$$

since in the relaxation time approximation $\boldsymbol{L}_k = \tau_k \boldsymbol{v}_k$. Now $-e\boldsymbol{E} \cdot \boldsymbol{L}_k$ is just the energy $\Delta\varepsilon$ acquired by the electron from the field $\boldsymbol{E}$ while travelling the distance $\boldsymbol{L}_k$, and in fact there is no real need to think in terms of $\boldsymbol{k}_{\mathrm{d}}$; since f_0 depends only on ε, we can rewrite (9.3) as

$$f(\boldsymbol{k}) = f_0(\varepsilon_k - \Delta\varepsilon) \tag{9.5}$$

and this leads immediately to (9.4).

Moreover it is clear that the result (9.4) is not restricted to the relaxation time approximation. In section (8.3) we saw how to derive a general expression for the vector mean free path $\boldsymbol{L}_k$ – the average distance travelled by an electron, starting at time t in state $\boldsymbol{k}$, before its motion becomes completely randomized. Now there is a complete symmetry (if $\boldsymbol{B} = 0$) between the average motion of the electron *after* time t and *before* time t; if it travels on average a distance $\boldsymbol{L}_k$ after time t before its motion becomes completely randomized, it will also on average have travelled the same distance before time t, if we trace its history back in time. The energy picked up by the electron from the field $\boldsymbol{E}$ will therefore be, on average, $-e\boldsymbol{E} \cdot \boldsymbol{L}_k$, and (9.5) again leads to (9.4) – not just in the relaxation time approximation, but quite generally.

From (9.2) and (9.4), the current density $\boldsymbol{J}$ produced by a field $\boldsymbol{E}$ is given by

$$\boldsymbol{J} = -(e^2/4\pi^3) \int \boldsymbol{v}_k (\boldsymbol{E} \cdot \boldsymbol{L}_k) \frac{\partial f_0}{\partial \varepsilon} \mathrm{d}^3 \boldsymbol{k} \tag{9.6}$$

This result shows that $\boldsymbol{J}$ is determined entirely by the electrons at and near the FS, since $\partial f_0/\partial \varepsilon$ vanishes except for ε close to ε_{F}. We saw in section (3.1) that $\boldsymbol{J}$ will not necessarily be parallel to $\boldsymbol{E}$ unless the crystal is cubic; in general, $\boldsymbol{J}$ and $\boldsymbol{E}$ are related by a *tensor* σ_{ij} such that

$$J_i = \sum_j \sigma_{ij} E_j \quad (i, j = x, y \text{ or } z) \tag{9.7}$$

From (9.6) we have

$$J_i = -(e^2/4\pi^3) \int v_{ki} (\boldsymbol{E} \cdot \boldsymbol{L}_k) \frac{\partial f_0}{\partial \varepsilon} \mathrm{d}^3 \boldsymbol{k},$$

so that

$$\sigma_{ij} = -(e^2/4\pi^3) \int v_{ki} \boldsymbol{L}_{kj} \frac{\partial f_0}{\partial \varepsilon} \mathrm{d}^3 \boldsymbol{k} \tag{9.8}$$

If we define $\bar{\sigma} = \frac{1}{3}(\sigma_{xx} + \sigma_{yy} + \sigma_{zz})$ as the average of the conductivity components σ_{ii} in any three perpendicular directions, it follows that

$$\bar{\sigma} = -(e^2/12\pi^3) \int \boldsymbol{v}_k \cdot \boldsymbol{L}_k \frac{\partial f_0}{\partial \varepsilon} \mathrm{d}^3 \boldsymbol{k} \tag{9.9}$$

(For a cubic metal, symmetry requires that $\sigma_{xx} = \sigma_{yy} = \sigma_{zz} = \bar{\sigma}$, and all other σ_{ij} components vanish. We then have $\boldsymbol{J} = \bar{\sigma} \boldsymbol{E}$ for all directions of $\boldsymbol{E}$, as expected.) From now on, we shall write $\bar{\sigma}$ as σ, and take it as our measure of the electrical conductivity.

By writing $\mathrm{d}^3\boldsymbol{k} = \mathrm{d}k_\perp \mathrm{d}S_{\boldsymbol{k}} = (\mathrm{d}k_\perp/\mathrm{d}\varepsilon)\mathrm{d}\varepsilon \mathrm{d}S_{\boldsymbol{k}} = \mathrm{d}\varepsilon \mathrm{d}S_{\boldsymbol{k}}/\hbar v_{\boldsymbol{k}}$, we can rewrite (9.9) in terms of a surface integral:

$$\begin{aligned} \sigma &= -(e^2/12\pi^3) \int \frac{\partial f_0}{\partial \varepsilon} \mathrm{d}\varepsilon \int_\varepsilon \mathrm{d}S_k (\boldsymbol{v}_k \cdot \boldsymbol{L}_k)/\hbar v_k \\ &= -\int \frac{\partial f_0}{\partial \varepsilon} \sigma(\varepsilon)\, \mathrm{d}\varepsilon \end{aligned} \tag{9.10}$$

where

$$\begin{aligned} \sigma(\varepsilon) &= (e^2/12\pi^3) \int_\varepsilon \mathrm{d}S_{\boldsymbol{k}} (\boldsymbol{v}_k \cdot \boldsymbol{L}_k) \hbar v_{\boldsymbol{k}} \\ &= (e^2/12\pi^3\hbar) \int_\varepsilon \mathrm{d}S_k L_{k,\parallel} \end{aligned} \tag{9.11}$$

Here $L_{k,\parallel}$ is the component of $\boldsymbol{L}_k$ parallel to $\boldsymbol{v}_k$, and the integral in (9.11) is over the constant-energy surface $S_k = S(\varepsilon)$ in k-space. Since $\sigma(\varepsilon)$ will usually vary only slightly over the energy range $\varepsilon_F \pm kT$ or so in which $\partial f_0/\partial\varepsilon$ is appreciable, we can usually take it outside the integral in (9.10), and write

$$\sigma = \sigma(\varepsilon_F) = (e^2/12\pi^3\hbar)\int_{\varepsilon_F} dS_k \boldsymbol{L}_{k,\parallel} \tag{9.12}$$

9.2 TEMPERATURE DEPENDENCE OF σ

Equation (9.12) is a very general expression for σ. For free electrons it reduces at once to the familiar result $\sigma = ne^2\tau/m$ (problem 9.1). For a real metal, despite the simple form of (9.12), it is a formidably difficult task to calculate σ from first principles, and only since about 1975 has it been possible to do so with reasonable accuracy. We shall confine ourselves here to a qualitative discussion, looking particularly at the expected temperature-dependence of σ.

Since it is difficult to solve (8.23) for $\boldsymbol{L}_k$, we shall assume that $\boldsymbol{L}_k$ is parallel to $\boldsymbol{v}_k$, so that we can write $L_{k,\parallel} = L_k = v_k\tau_k^e$, and (9.12) reduces to

$$\sigma = (e^2/12\pi^3\hbar)\int_{\varepsilon_F} v_k\tau_k^e dS_k. \tag{9.13}$$

We shall also use the approximation (8.25) for τ_k^e. Though this is a rather crude approximation, it very much simplifies the problem while still retaining the essential features. For scattering by point defects, with $P_{kk'}$ given by (8.5), we then have

$$1/\tau_k^e = \int dS_{k'}(n_i/4\pi^2\hbar^2 v_{k'})|V_{kk'}|^2(1-\cos\theta) \tag{9.14}$$

where $V_{kk'}$ varies with the scattering vector $\boldsymbol{k}_s = |\boldsymbol{k} - \boldsymbol{k}'|$ in the kind of way shown in Fig. 8.1. As k_s increases, so does the scattering angle θ; large-angle scattering ($\theta \sim \pi$) will typically correspond to $k_s r_0 \sim 4$–5, or $k_s r_1 \sim 1.5$–2, so that $|V_{kk'}|^2$ is considerably smaller for $\theta \sim \pi$ than for $\theta \sim 0$. Because of the $(1-\cos\theta)$ weighting factor, $1/\tau_k^e$ will therefore be smaller than the unweighted scattering rate $1/\tau_k^0$, typically by 30–40% or so. But nothing in (9.14) depends on temperature, so that the residual resistivity ρ_r due to defect scattering will likewise be temperature-independent.

We shall look at the evaluation of (9.14) in more detail in section 13.3; for now we simply note that it yields estimates of $1/\tau^e$ (and hence of ρ_r) which turn out to be too large by a factor of 4 or 5, because $V_s(\boldsymbol{r})$ (eqn (8.7)) is really too strong for the Born approximation to be valid.

For phonon scattering, unlike impurity scattering, $1/\tau^e$ will be strongly temperature-dependent, because of the factor n_q. Inserting (8.19) in (8.25), with $V_{kk'}$ given by (8.18), we find

$$1/\tau_k^e = \int dS_{k'}(2n_q/4\pi^2\hbar^2 v_{k'})(V_0^2\hbar q^2/2NM\omega_q)(1-\cos\theta)$$

$$= (V_0^2/4\pi^2\hbar NM)\int dS_{k'} n_q(q^2/\omega_q v_{k'})(1-\cos\theta) \qquad (9.15)$$

For $T \gtrsim \theta_D$, we shall have $\hbar\omega_q \lesssim kT$ for all the phonon modes, and (8.12) shows that we can then put $n_q \approx kT/\hbar\omega_q$ in (9.15). Since none of the other terms in (9.15) is temperature-dependent, we therefore have $1/\tau_k^e \propto T$, and so $\rho_t \propto T$, where ρ_t is the resistivity due to phonon scattering. This is precisely the temperature dependence usually observed experimentally at room temperature and above (Fig. 1.1), and moreover (9.15) also predicts the actual magnitude of τ^e reasonably well (problem 9.2).

The temperature dependence is very different for $T \ll \theta_D$. The only phonons which are then appreciably excited are those for which $\hbar\omega_q \lesssim kT$. For these low-frequency phonons we can put $\hbar\omega_q \approx \hbar q v_s$, where v_s is the velocity of sound, so that the condition $\hbar\omega_q \lesssim kT$ becomes $q \lesssim q_T$, where $\hbar q_T v_s = kT$. Now for N-processes, the scattering angle θ will be proportional to q if q is small (Fig. 9.1), so that for $q \sim q_T$, $(1-\cos\theta) \approx \frac{1}{2}\theta^2$ will be proportional to q_T^2 and hence to T^2. At the same time the area of FS, $\delta S_{k'}$, into which scattering can occur will be roughly πq_T^2 (Fig. 9.1), again proportional to T^2. Finally,

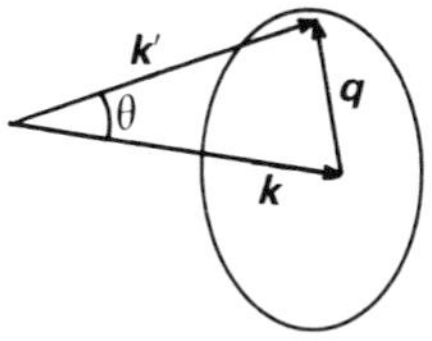

Fig. 9.1 If $\boldsymbol{q}$ is small, $\theta \propto q$, and scattering is into a disc on the FS of area πq^2.

$|V_{kk'}|^2$ itself contains the factor $q^2/\omega_q = q/v_s$, and will hence be proportional to T for $q \sim q_T$. Putting all these factors together, we find that $1/\tau^e$ (and hence ρ_t) should vary as T^5 at low temperatures, if U-processes are ignored. (This rather hand-waving derivation can easily be confirmed by writing $dS_{k'} \approx 2\pi q\,dq$, $(1 - \cos\theta) \approx q^2/2k_F^2$ and $q^2/\omega_q = q/v_s$ in (9.15). We then have an integral of the form $\int q^4 dq/(e^{\alpha q/T} - 1)$, $= (T/\alpha)^5 \int x^4 dx/(e^x - 1)$, where $\alpha = \hbar v_s/k_b$.) The result $\rho_t \propto T^5$ for $T \ll \theta$, was first derived by Bloch in 1928, and is sometimes called Bloch's T^5 'law'.

Experimentally, ρ_t does indeed vary with T very rapidly at low temperatures, but not always as T^5. If we try to fit the low-temperature behaviour by writing $\rho_t \propto T^n$, n may be anywhere between 4 and 8 or more, and may itself vary with T. This is not too surprising, because the T^5 'law' ignores U-processes, and they can have a drastic effect on ρ_t. In U-processes (Fig. 8.3), a low-$\boldsymbol{q}$ phonon can produce large-angle scattering, for which $1 - \cos\theta \sim 2$. Clearly, if $1 - \cos\theta$ for N-processes is vanishing as T^2, U-processes can easily dominate (9.15), even if n_q for these processes is small.

The relative importance of N-processes and U-processes, and the consequent form of $\rho_t(T)$, depend very much on the detailed shape of the FS. Figure 9.2 shows two possible examples. In Fig. 9.2(a) (as in Fig. 8.3, and as in the alkali metals) the FS comes close to the BZ boundary but does not touch it, so that only phonons with $q \geqslant q_{min}$ can cause U-processes. These phonons will have energy of at least

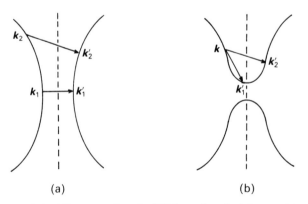

Fig. 9.2 (a) If the FS approaches the BZ boundary but does not touch it, the shortest $\boldsymbol{q}$ vectors that can produce U-processes from $\boldsymbol{k}_1$ and $\boldsymbol{k}_2$ are as shown. (b) If the FS touches the zone boundary, there is no sharp distinction between N-processes and U-processes.

$\hbar\omega(\boldsymbol{q}_{\min})$, and their number n_q will therefore fall exponentially at low temperatures. The contribution from U-process scattering to $1/\tau_k^e$ will therefore also fall off exponentially, from (9.15). As shown in Fig. 9.2(a), the size of $q_{\min}$ (and hence the phonon energy) will depend on the position of $\boldsymbol{k}$ on the FS, and $1/\tau_k^e$ and its temperature dependence will therefore vary over the FS. Eventually, when $kT \ll \hbar\omega(\boldsymbol{q}_{\min})$ for all $\boldsymbol{k}$, the contribution of U-processes to $1/\tau_k^e$ should become negligible, but in practice this only happens at very low temperatures – in the alkalis, only below 1K.

Except in the alkalis, the FS touches the BZ boundary at some point, and Fig. 9.2(b) for example shows schematically the resultant 'neck' in the (111) direction in Cu (cf. Fig. 5.7). There is now no $q_{\min}$, and no sharp distinction between N-processes and U-processes. Formally, scattering from $\boldsymbol{k}$ to $\boldsymbol{k}_1'$ is an N-process and from $\boldsymbol{k}$ to $\boldsymbol{k}_2'$ is a U-process, but (as in Fig. 8.3b)) there is no abrupt change in the

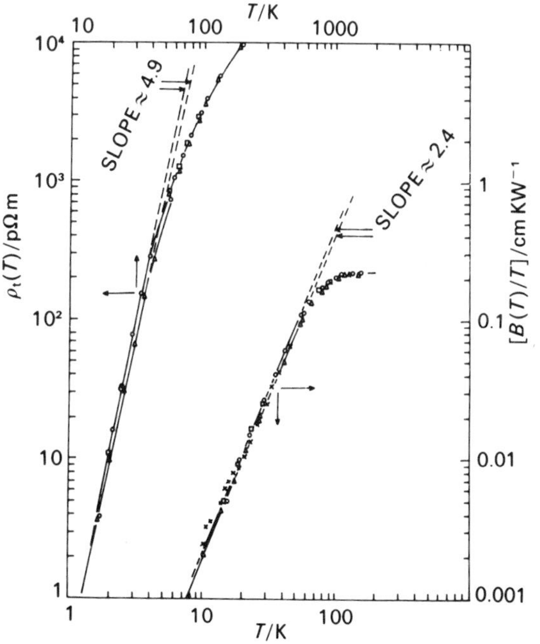

Fig. 9.3 Log-log plots of $\rho_t(T)$ and $B(T)/T$ (see equation (10.11)) against T for three samples of Cu having $\rho_r = 4.6 \times 10^{-11}$, 5.1×10^{-10} and $5.8 \times 10^{-10}\,\Omega$m, showing that at low temperatures ρ_t rises as $T^{4.9}$ and $B(T)/T$ as $T^{2.4}$. [From White (1953), *Aust. J. Phys.* **6**, 397.]

physics as we cross the zone boundary. In this example, we might expect $1/\tau_k^e$ to follow something like the T^5 variation of an N-process rather than the exponential fall-off of a U-process with a finite $q_{\min}$, but the precise temperature dependence will depend on the detailed shape of the neck region.

Without detailed calculation, then, the only general statement we can make about the low-temperature behaviour of ρ_t is that it is likely to vary rapidly with T – probably as T^5 or faster. This is indeed what is observed experimentally, though as shown in Fig. 9.3 the variation is sometimes slightly slower than T^5.

9.3 MATTHIESSEN'S RULE

So far we have discussed ρ_r and ρ_t separately. What happens when defect scattering and phonon scattering are both present? We can reasonably assume that the basic scattering probabilities $P_{kk'}$ for the two processes are simply additive, as long as the defects are not too numerous. (If they become too numerous – as in a concentrated alloy, for example – they will affect both $\varepsilon(\boldsymbol{k})$ and $\omega(\boldsymbol{q})$, and the theory becomes a lot more complicated.) It follows that the total scattering rate $1/\tau_k^0$ of (8.20) is just the sum of contributions from thermal scattering by phonons and residual scattering by static defects:

$$1/\tau_k^0 = 1/\tau_{k,t}^0 + 1/\tau_{k,r}^0 \tag{9.16}$$

It is not so easy to see from (8.23) how the vector mean free paths $\boldsymbol{L}_k$ due to phonon scattering and defect scattering will combine, or to see from (8.24) how the effective scattering rates $1/\tau_k^e$ will combine, unless we again use the rather crude approximation (8.25) in place of (8.24). If we do use that approximation, we at once have a result for $1/\tau^e$ similar to (9.16):

$$1/\tau_k^e = 1/\tau_{k,t}^e + 1/\tau_{k,r}^e \tag{9.17}$$

But even if (9.17) is valid, Matthiessen's rule (2.11) will only hold exactly if the ratio $\tau_{k,t}^e/\tau_{k,r}^e$ is the same at all points on the FS; if $\tau_{k,r}^e$ varies from point to point, $\tau_{k,t}^e$ must vary in just the same way (problem 9.3). It is unlikely that this condition is ever satisfied in a real metal, so it is not surprising that Matthiessen's rule is not obeyed exactly. Nevertheless, it is usually rather a good approximation (problem 9.4): if $\rho(T)$ is measured on a number of samples of a given metal, of different purity, the results can usually be rather well represented by writing $\rho = \rho_r + \rho_t(T)$, where the residual resistivity ρ_r varies from sample to sample but the temperature-dependent part $\rho_t(T)$ is the same for all samples, and vanishes at $T = 0$ (Figs 1.1 and 9.3).

9.4 THE KONDO EFFECT

Whether or not Matthiessen's rule is obeyed exactly, we would expect the general form of $\rho(T)$ to be as shown in Fig. 1.1; phonon scattering should die away to zero as $T \to 0$, leaving a temperature-independent residual resistivity ρ_r due to static defects. For most metal samples, this is precisely what happens, but it has been known since about 1930 that in some samples $\rho(T)$ goes through a shallow minimum ρ_{min} as the temperature is lowered, typically at $T \sim 20\,K$, and then rises again, typically by about 10% of ρ_{min}, before flattening out as $T \to 0$. It was established eventually that this resistance minimum effect – along with related anomalies in the thermoelectric power and in the magnetoresistance – was due to small amounts of magnetic impurity, and Fig. 9.4 shows typical $\rho(T)$ curves for 0.05, 0.1 and 0.2 atomic % Fe in Cu.

This effect remained unexplained until 1964, when Kondo showed that impurity scattering by a magnetic ion could indeed become stronger as the temperature fell. The effect is a very subtle one, and we can only give a rough idea of it here.

If we calculate the scattering from magnetic ions using (8.1) and (8.2), i.e. using the Born approximation, it turns out to be temperature-independent, just like the scattering from any other static defect. But the Born approximation is in fact only the first term in a full perturbation-theory treatment of scattering. In most scattering problems, nothing qualitatively new emerges if one goes to the next term – the second Born approximation – but Kondo showed that this particular problem was different. In the second Born approximation, the scattering from a magnetic ion rises as T falls, and this explains the Kondo effect, as the resistance minimum effect is now called.

In (8.1), the term $|V_{\boldsymbol{kk}'}|^2$ represents direct scattering from state $\boldsymbol{k}$ to state $\boldsymbol{k}'$. In the second Born approximation, this term is replaced by

$$\left| V_{\boldsymbol{kk}'} + \sum_{\boldsymbol{k}''} \frac{V_{\boldsymbol{kk}''} V_{\boldsymbol{k}''\boldsymbol{k}'}}{\varepsilon_{\boldsymbol{k}} - \varepsilon_{\boldsymbol{k}''}} \right|^2 \tag{9.18}$$

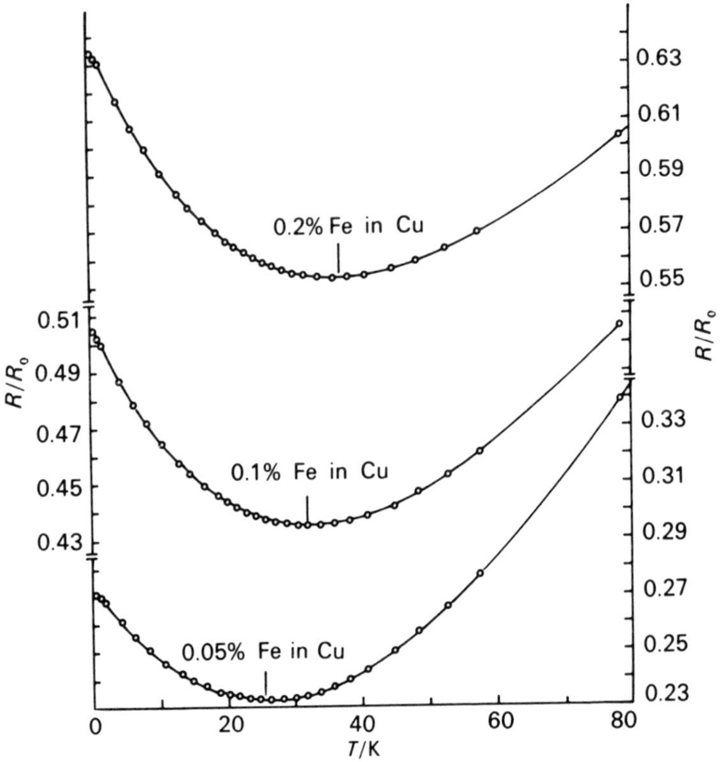

Fig. 9.4 The resistance minimum or Kondo effect in three Cu samples. [From Franck *et al.* (1961), *Proc. Roy. Soc.*, **A263**, 494.]

where the second part represents scattering from $\boldsymbol{k}$ to $\boldsymbol{k}'$ via the intermediate state $\boldsymbol{k}''$, which differs in energy from $\boldsymbol{k}$ and $\boldsymbol{k}'$. There are various terms of this kind, depending on the spin orientations of $\boldsymbol{k}$, $\boldsymbol{k}'$ and $\boldsymbol{k}''$, and the interesting terms – the ones that lead to temperature-dependent scattering – are those in which the intermediate state $\boldsymbol{k}''$ is of opposite spin to $\boldsymbol{k}$ and $\boldsymbol{k}'$. (The change in electron spin from $\uparrow$ to $\downarrow$ to $\uparrow$ must be balanced by opposite changes in the spin on the ion – that is why magnetic ions are needed to produce the Kondo effect.) Now this indirect transition from $\boldsymbol{k}$ to $\boldsymbol{k}'$ can occur in two ways. In the first, a $\boldsymbol{k}\uparrow$ electron is scattered first into an empty $\boldsymbol{k}''\downarrow$ state, and then into a $\boldsymbol{k}'\uparrow$ state. In the second, an electron already in a $\boldsymbol{k}''\downarrow$ state is first scattered into $\boldsymbol{k}'\uparrow$, and then the $\boldsymbol{k}\uparrow$ electron is scattered into the vacated $\boldsymbol{k}''\downarrow$ state. The first involves a factor $1 - f_{\boldsymbol{k}''\downarrow}$, the probability that $\boldsymbol{k}''\downarrow$ is initially empty, and the second involves $f_{k''\downarrow}$, the probability that it is initially full. The second term in (9.18) thus becomes, more precisely,

$$\sum_{k''}[V_{kk''}V_{k''k'}(1-f_{k''\downarrow})+V_{k''k'}V_{kk''}f_{k''\downarrow}]/(\varepsilon_k-\varepsilon_{k''})$$

Now if the scattering processes did not involve a spin flip, the two products $V_{kk''}V_{k''k'}$ and $V_{k''k'}V_{kk''}$ would be equal, so that the $f_{k''\downarrow}$ factors would cancel, and we should be left, as usual, with a temperature-independent scattering process. But for spin-flip scattering, the matrix elements $V_{kk''}$, etc. are more subtle objects involving spin operators, which no longer commute, and the $f_{k''\downarrow}$ factors no longer cancel. We are thus left with a term involving, essentially, the quantity

$$I=\sum_{k''}f_{k''\downarrow}/(\varepsilon_k-\varepsilon_{k''}) \tag{9.19}$$

summed over all intermediate states $\boldsymbol{k}''$. It is this term which, through the temperature-dependent factor $f_{k''}$, leads to temperature-dependent scattering. With negligible error we can replace $f_{k''\downarrow}$ by $f_{0k''}=f_0(\varepsilon_{k''})$, and we can then replace the sum in (9.19) by an integral over energy:

$$I=\frac{1}{2}\int_{\varepsilon_{\min}}^{\infty}g(\varepsilon)f_0(\varepsilon)\,\mathrm{d}\varepsilon/(\varepsilon_k-\varepsilon) \tag{9.20}$$

where $\varepsilon_{\min}$ is the energy at the bottom of the conduction band and $g(\varepsilon)$ is the density of states (7.2), so that $\frac{1}{2}g(\varepsilon)\delta\varepsilon$ is the number of up-spin (or down-spin) states in the energy range between ε and $\varepsilon+\delta\varepsilon$. We can evaluate (9.20), for $\varepsilon_k=\varepsilon_F$, by a simple trick. From (1.11), it follows at once that

$$\frac{\partial f_0}{\partial T}=-\frac{\varepsilon-\varepsilon_F}{T}\frac{\partial f_0}{\partial\varepsilon} \tag{9.21}$$

(neglecting a very small term involving $\partial\varepsilon_F/\partial T$), so that for $\varepsilon_k=\varepsilon_F$,

$$\begin{aligned}\frac{\partial I}{\partial T}&=\frac{1}{2T}\int_{\varepsilon_{\min}}^{\infty}g(\varepsilon)\frac{\partial f_0}{\partial\varepsilon}\,\mathrm{d}\varepsilon\\&=-g(\varepsilon_F)/2T\end{aligned}$$

Integrating back again with respect to T, we find

$$I = \tfrac{1}{2}g(\varepsilon_F)\ln(\varepsilon_0/kT) \tag{9.22}$$

where ε_0 is a constant of integration. More detailed calculation shows that $\varepsilon_0 \approx \varepsilon_F - \varepsilon_{min}$, and also shows that I is still given approximately by (9.22) when $\varepsilon_k \neq \varepsilon_F$ if T is replaced by $T\sqrt{(1+\eta^2)}$, with $\eta = (\varepsilon_k - \varepsilon_F)/kT$. The important result, though, is that I varies as $\ln(1/T)$, so that the resultant term in the scattering rate, and in the resistivity, likewise varies as $\ln(1/T)$. This is in good agreement with the observed behaviour below the resistance minimum (Fig. 9.4). At low enough temperatures, however, ρ ceases to rise logarithmically and tends to a constant value. Magnetic susceptibility measurements show that this is because the magnetic ion, when immersed in a sea of conduction electrons, effectively loses its magnetic moment at low enough temperatures. This slow transition to a non-magnetic ground state proved remarkably difficult to handle theoretically, and a satisfactory analytic solution to the 'Kondo problem' was found only in 1980. In effect, the moment of the ion becomes screened by the conduction electrons, but the detailed theory is far too subtle to discuss here.

10

Metals in a temperature gradient

10.1 THERMAL CONDUCTIVITY

An electric current is a flow of charge; a heat current is a flow of thermal energy. An electron moving with velocity $\boldsymbol{v}$ contributes $-e\boldsymbol{v}$ to the electric current, and it contributes $\varepsilon_T \boldsymbol{v}$ to the heat current, where ε_T is the thermal energy of the electron. But what *is* the 'thermal energy' of an electron? Clearly it is not just the energy $\varepsilon_{\boldsymbol{k}}$: for one thing, $\varepsilon_{\boldsymbol{k}}$ is not uniquely defined, because it includes a contribution from the potential energy $V(\boldsymbol{r})$, which (like all potential energies) is measured from some arbitrary zero; for another thing, $\varepsilon_{\boldsymbol{k}}$ certainly does not vanish at $T=0$, whereas we would expect the thermal energy of the relevant electrons – the electrons at the FS – to vanish at $T=0$. By considering the relation between heat flow and entropy flow, it can be shown that the correct measure of the thermal energy is $\varepsilon_{\boldsymbol{k}} - \varepsilon_F$, which makes sense; this quantity is independent of the choice of potential energy zero, and it vanishes for electrons within $\pm kT$ of the FS as $T \to 0$. Corresponding to equation (9.2) for the electric current density $\boldsymbol{J}$, we thus have the equation

$$\boldsymbol{Q}(\boldsymbol{r}) = (1/4\pi^3) \int (\varepsilon_{\boldsymbol{k}} - \varepsilon_F)\, \boldsymbol{v}_{\boldsymbol{k}} f(\boldsymbol{k}, \boldsymbol{r})\, d^3\boldsymbol{k} \tag{10.1}$$

for the heat current density $\boldsymbol{Q}$. (In principle we should write $\varepsilon_F(\boldsymbol{r})$ here, since ε_F depends slightly on T and hence on $\boldsymbol{r}$ if $T = T(\boldsymbol{r})$, but the effect of this is very small and, as in (9.21), we shall ignore it.)

If $f(\boldsymbol{k}, \boldsymbol{r}) = f_0(\boldsymbol{k}, \boldsymbol{r})$, then $\boldsymbol{Q} = 0$; in thermal equilibrium, $\boldsymbol{Q}$ vanishes, like $\boldsymbol{J}$, and for the same reasons. But a temperature gradient

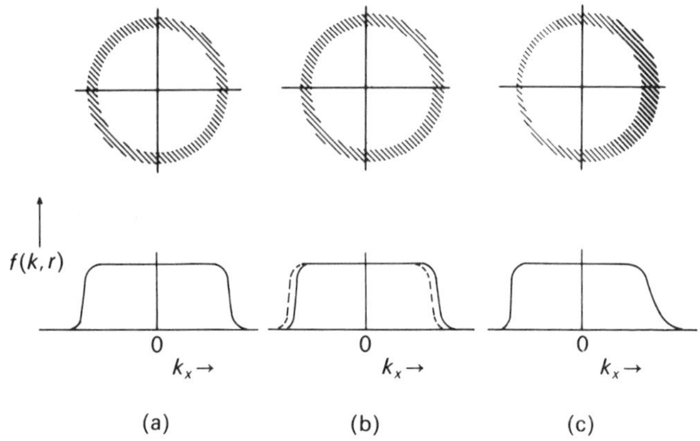

Fig. 10.1 The blurred Fermi surface and the form of $f(\boldsymbol{k},\boldsymbol{r})$, (a) with no applied fields; (b) when the whole FS is displaced by a field E_x; (c) when one side of the FS is hotter and therefore more blurred than the other, because of a temperature gradient G_x.

$\boldsymbol{G} = -\nabla T$ will cause $f(\boldsymbol{k},\boldsymbol{r})$ to differ from the equilibrium value $f_0(\boldsymbol{k},\boldsymbol{r})$ – that is, the equilibrium value appropriate to the local temperature T at the point $\boldsymbol{r}$ – and will hence produce a heat current. If the temperature is higher to the left of point $\boldsymbol{r}$ than to the right, electrons arriving from the left will be 'hotter' than those arriving from the right, so that the FS at point $\boldsymbol{r}$ will be more blurred on one side than the other. The result is to perturb $f(\boldsymbol{k},\boldsymbol{r})$ away from $f_0(\boldsymbol{k},\boldsymbol{r})$ in the way shown schematically in Fig. 10.1c. Clearly this perturbation is quite different from that produced by an electric field $\boldsymbol{E}$, and shown in Fig. 10.1b.

To find the actual form of $f(\boldsymbol{k},\boldsymbol{r})$, suppose first, as we did in section 9.1, that the relaxation time approximation holds: that the electron starts out from each collision with the unperturbed distribution function f_0, and that at any instant it has on average travelled for a time τ_k and for a distance $\boldsymbol{L}_k^T = \tau_k \boldsymbol{v}_k$ since its last collision. It will therefore have come from a place where the relevant value of f_0 was $f_0(\boldsymbol{k},\boldsymbol{r}-\boldsymbol{L}_k^T)$, and we have

$$f(\boldsymbol{k},\boldsymbol{r}) = f_0(\boldsymbol{k},\boldsymbol{r}-\boldsymbol{L}_k^T) \tag{10.2}$$

as the equivalent of (9.3). Just as in deriving (9.4) from (9.3), we can now write

$$
\begin{aligned}
f(\boldsymbol{k},\boldsymbol{r}) &= f_0(\boldsymbol{k},\boldsymbol{r}) - \boldsymbol{L}_k^T\cdot\frac{\partial f_0}{\partial \boldsymbol{r}} \\
&= f_0(\boldsymbol{k},\boldsymbol{r}) - \boldsymbol{L}_k^T\cdot\frac{\mathrm{d}T}{\mathrm{d}\boldsymbol{r}}\frac{\partial f_0}{\partial T} \\
&= f_0(\boldsymbol{k},\boldsymbol{r}) - \boldsymbol{L}_k^T\cdot\boldsymbol{G}(\varepsilon_{\boldsymbol{k}} - \varepsilon_{\mathrm{F}})T^{-1}\frac{\partial f_0}{\partial \varepsilon}
\end{aligned} \tag{10.3}
$$

writing $\boldsymbol{G} = -\mathrm{d}T/\mathrm{d}\boldsymbol{r}$ and using (9.21). Just as with (9.4), it is clear that the result (10.3) is not restricted to the relaxation time approximation; it remains true much more generally, if we interpret $\boldsymbol{L}_k^T$ as an appropriate vector mean free path. But except in the relaxation time approximation, $\boldsymbol{L}_k^T$ is not necessarily the same as the vector mean free path of (8.23) and (9.4), which is why we have called it $\boldsymbol{L}_k^T$ rather than $\boldsymbol{L}_{\boldsymbol{k}}$. We shall come back to this point later, but first let us complete the calculation of the thermal conductivity κ, by inserting (10.3) in (10.1). We thus find

$$
\boldsymbol{Q} = -(1/4\pi^3 T)\int \boldsymbol{v}_{\boldsymbol{k}}(\varepsilon_{\boldsymbol{k}} - \varepsilon_{\mathrm{F}})^2\boldsymbol{G}\cdot\boldsymbol{L}_k^T\frac{\partial f_0}{\partial \varepsilon}\,\mathrm{d}^3\boldsymbol{k} \tag{10.4}
$$

Following exactly the same path that we used to get from (9.6) to (9.10) and (9.11), we can write this in the form $Q_i = \sum_j \kappa_{ij} G_j$, where

$$
\kappa_{ij} = -(1/4\pi^3 T)\int \boldsymbol{v}_{\boldsymbol{k}i}\boldsymbol{L}_{kj}^T(\varepsilon_{\boldsymbol{k}} - \varepsilon_{\mathrm{F}})^2\frac{\partial f_0}{\partial \varepsilon}\,\mathrm{d}^3\boldsymbol{k}, \tag{10.5}
$$

so that the average of κ_{xx}, κ_{yy} and κ_{zz} (which we call κ from now on) is

$$
\begin{aligned}
\kappa &= -(1/12\pi^3 T)\int \boldsymbol{v}_{\boldsymbol{k}}\cdot\boldsymbol{L}_k^T(\varepsilon_{\boldsymbol{k}} - \varepsilon_{\mathrm{F}})^2\frac{\partial f_0}{\partial \varepsilon}\,\mathrm{d}^3\boldsymbol{k} \\
&= -(1/12\pi^3 T)\int \boldsymbol{v}_{\boldsymbol{k}}\cdot\boldsymbol{L}_k^T(\varepsilon_{\boldsymbol{k}} - \varepsilon_{\mathrm{F}})^2\frac{\partial f_0}{\partial \varepsilon}\,\mathrm{d}\varepsilon\frac{\mathrm{d}\boldsymbol{k}_\perp}{\mathrm{d}\varepsilon}\,\mathrm{d}S_{\boldsymbol{k}}
\end{aligned} \tag{10.6}
$$

Now $(\varepsilon - \varepsilon_{\mathrm{F}})^2(\partial f_0/\partial\varepsilon)$, like $\partial f_0/\partial\varepsilon$ itself, falls exponentially to zero for $|\varepsilon - \varepsilon_{\mathrm{F}}| \gg kT$, and in fact

$$
-\int_{-\infty}^{\infty}(\varepsilon - \varepsilon_{\mathrm{F}})^2\frac{\partial f_0}{\partial \varepsilon}\,\mathrm{d}\varepsilon = \pi^2 k^2 T^2/3 \tag{10.7}
$$

(problem 10.1), so that corresponding to (9.10) and (9.11) we can write

$$\kappa = -\frac{3}{\pi^2 k^2 T^2}\int_{-\infty}^{\infty}(\varepsilon-\varepsilon_{\mathrm{F}})^2\frac{\partial f_0}{\partial\varepsilon}\,\mathrm{d}\varepsilon\,\kappa(\varepsilon) \tag{10.8}$$

where

$$\begin{aligned}\kappa(\varepsilon) &= \frac{\pi^2 k^2 T}{3}\frac{1}{12\pi^3}\int_{\varepsilon}\mathrm{d}S_{\boldsymbol{k}}\boldsymbol{v}_{\boldsymbol{k}}\cdot\boldsymbol{L}_{\boldsymbol{k}}^{T}/\hbar v_{\boldsymbol{k}}\\ &= \frac{\pi^2 k^2 T}{3}\frac{1}{12\pi^3\hbar}\int_{\varepsilon}\mathrm{d}S_{\boldsymbol{k}}L_{\boldsymbol{k},\parallel}^{T}\end{aligned} \tag{10.9}$$

If $\kappa(\varepsilon)$ does not vary significantly over the range $\varepsilon_{\mathrm{F}} \pm kT$, we can take it outside the integral in (10.8), and from (10.7) we then have $\kappa = \kappa(\varepsilon_{\mathrm{F}})$ as our final expression for the mean thermal conductivity. This is very similar to the expression (9.12) for the mean electrical conductivity σ. Combining these two expressions, the Lorenz number $\mathscr{L} = \kappa/\sigma T$ is given by

$$\mathscr{L} = (\pi^2 k^2/3e^2)\int_{\varepsilon_{\mathrm{F}}}\mathrm{d}S_{\boldsymbol{k}}L_{\boldsymbol{k},\parallel}^{T}\bigg/\int_{\varepsilon_{\mathrm{F}}}\mathrm{d}S_{\boldsymbol{k}}L_{\boldsymbol{k},\parallel} \tag{10.10}$$

Thus if the thermal and electrical vector mean free paths $\boldsymbol{L}^T$ and $\boldsymbol{L}$ are the same, we at once recover the Sommerfeld expression (2.10) for the Lorenz number, but now our derivation is far more general.

As we saw in Chapter 2, the Sommerfeld expression agrees very well with experiment at room temperature and above, and also at very low temperatures, but not at intermediate temperatures (Fig. 1.3), and we can now understand why, in terms of (10.10). The electrical mfp $\boldsymbol{L}_{\boldsymbol{k}}$ of (8.23) and (9.4) is the average distance an electron travels after time t before its direction of motion becomes completely randomized. By symmetry it will have travelled the same distance *before* time t (if $\boldsymbol{B} = 0$), so that $-e\boldsymbol{E}\cdot\boldsymbol{L}_{\boldsymbol{k}}$ in (9.4) correctly represents the energy it has picked up from the field $\boldsymbol{E}$. But the thermal mfp in (10.2) and (10.3) has a different significance; it represents the distance the electron needs to move after time t before it comes into thermal equilibrium with its surroundings, or the distance it has travelled, before time t, since being in thermal equilibrium. Roughly speaking, $\boldsymbol{L}_k$ measures how far the electron goes before it forgets where it was going, and $\boldsymbol{L}_k^T$ measures how far it goes before it forgets how hot it was, and the two distances are not necessarily the same.

To restore the distribution shown in Fig. 10.1c to that shown in Fig. 10.1a, collisions merely have to scatter electrons from just below the FS to just above it, or vice versa; there is no need for them to be scattered through any appreciable distance in $\boldsymbol{k}$-space. To restore Fig. 10.1b to Fig. 10.1a, by contrast, the collisions have to scatter the electrons all the way from one side of the FS to the other – a much greater distance. In other words, small-angle scattering, which has very little effect in limiting the electrical mfp $\boldsymbol{L}$, can be very effective in limiting the thermal mfp $\boldsymbol{L}^T$, provided that it is *inelastic*, so that the electron can gain or lose energy from its surroundings at each collision. If the scattering is elastic, it is no more effective in limiting $\boldsymbol{L}^T$ than $\boldsymbol{L}$; the only way that elastic collisions can restore Fig. 10.1c to Fig. 10.1a is by scattering electrons from one side of the FS to the other, exactly as for Fig. 10.1b.

At very low temperatures, then, when virtually all the scattering is elastic scattering by static defects, we expect $\boldsymbol{L}^T = \boldsymbol{L}$, and from (10.10) we therefore expect the Lorenz number $\mathscr{L}$ to tend to the Sommerfeld value $\pi^2k^2/3e^2$. We expect the same thing to happen at high temperatures, in fact, even though the scattering is then mainly inelastic phonon scattering, because for $T \gtrsim \theta_D$, the energy change at each collision will be small compared with kT, and will not do much to restore thermal equilibrium. Moreover, small-angle collisions are no longer very important for $T \gtrsim \theta_D$; most collisions will then scatter the electrons through large angles, and will be equally effective in restoring equilibrium whether they are elastic or not.

It is at moderately low temperatures, then, where $T < \theta_D$ but where most of the scattering is still by phonons rather than defects, that we expect the greatest difference between $\boldsymbol{L}^T$ and $\boldsymbol{L}$. To get an idea how big the difference might be, we can assume very roughly that as far as thermal conductivity is concerned, every collision effectively restores the electron to the equilibrium distribution, so that the relevant scattering rate is $1/\tau_k^0$, as given by (8.20). (The effective scattering rate is discussed more fully in Appendix B, but this approximation is good enough for our present purposes.) Now $1/\tau_k^0$ differs from the effective scattering rate for electrical conductivity,

$1/\tau_k^e$, by the omission of the $(1-\cos\theta)$ weighting factor (compare (8.20) and (8.25)), precisely as we might expect. This weighting factor represented the ineffectiveness of small-angle collisions, and we are now saying that as far as thermal conductivity is concerned they are *not* ineffective, so long as they are inelastic. Correspondingly, the $(1-\cos\theta)$ factor should now be omitted from (9.15), and it follows at once that at low temperatures $1/\tau_k^0$ should vary as T^3 rather than T^5. In the electrical case, we saw that the T^5 'law' was of doubtful validity, because it neglected U-processes, but in the thermal case, U-processes are much less important – simply because small-angle N-processes are much *more* important – and we can have a good deal more confidence in the T^3 result.

It follows from all this that at moderately low temperatures $\boldsymbol{L}^T$ may be much less than $\boldsymbol{L}$, although at still lower temperatures where defect scattering is dominant the two should become equal again, as they also should for $T > \theta_D$. From equation (10.10), we therefore expect the Lorenz number to vary with T in precisely the way it does experimentally, as shown in Fig. 1.3.

When both phonon scattering and defect scattering are present together, we can expect the thermal resistivity $W = 1/\kappa$ to obey a rule closely similar to Matthiessen's rule (section 9.3), with the same limitations. From (10.9), it is not W but WT which is proportional to $1/\boldsymbol{L}^T$, and hence to $1/\tau^T$ if $\boldsymbol{L}_k^T = \boldsymbol{v}_k \tau_k^T$. Assuming as in (9.17) that the total effective scattering rate $1/\tau_k^T$ is the sum of separate contributions from phonon scattering and from defect scattering, we thus expect to find

$$WT = A + B(T) \qquad (10.11)$$

as the analogue of Matthiessen's rule $\rho = \rho_r + \rho_t(T)$. Here A is the temperature-independent defect scattering term, and $B(T)$ is the phonon scattering term, which should vary as T^3 at low temperatures and as T at high temperatures. Since the ratio $\mathscr{L} = \kappa/\sigma T = \rho/WT$ should tend to the Sommerfeld value $\mathscr{L}_0$ at very low temperatures and at high temperatures, we expect to have $A = \rho_r/\mathscr{L}_0$, and at high temperatures we expect W to tend to the constant value $\rho/\mathscr{L}_0 T$.

Figure 9.3 shows the phonon scattering contribution to W for three Cu samples with very different values of A. The fact that they all give closely similar curves confirms that $B(T)$ is independent of A, as implied by (10.11), though the slopes of the curves are a little different from the predicted T^2 variation of $B(T)/T$.

10.2 THERMOELECTRIC EFFECTS

So far we have considered the electric current $\boldsymbol{J}$ produced by a field $\boldsymbol{E}$, and the heat current $\boldsymbol{Q}$ produced by a temperature gradient $\boldsymbol{G}$. In fact a field $\boldsymbol{E}$ will also produce a small heat current $\boldsymbol{Q}$ even if $\boldsymbol{G}=0$, and likewise $\boldsymbol{G}$ will produce a small electric current $\boldsymbol{J}$ even if $\boldsymbol{E}=0$. These small thermoelectric currents can be in the same direction as the main current or in the opposite direction, and arise from a slight inequality between the currents carried by 'hot' electrons (with $\varepsilon > \varepsilon_{\mathrm{F}}$) and 'cool' ones (with $\varepsilon < \varepsilon_{\mathrm{F}}$). For example, the heat current $\boldsymbol{Q}$ produced by $\boldsymbol{G}$ consists of a flow of hot electrons in one direction and of cool ones in the opposite direction; if the rates of flow in the two directions are not quite equal, there will be a net electric current $\boldsymbol{J}$. Likewise, the current $\boldsymbol{J}$ produced by $\boldsymbol{E}$ will include both hot and cool electrons flowing in the same direction, and if the flow includes more (or less) hot electrons than cool ones, there will be a net heat current $\boldsymbol{Q}$ one way or the other.

It is a simple matter to work out these currents. If we take the expression (10.3) for the distribution $f(\boldsymbol{k},\boldsymbol{r})$ produced by $\boldsymbol{G}$, and use it in (9.2), we find

$$\boldsymbol{J}=(e/4\pi^3 T)\int \boldsymbol{v}_{\boldsymbol{k}}(\boldsymbol{L}_{\boldsymbol{k}}^T\cdot\boldsymbol{G})(\varepsilon_{\boldsymbol{k}}-\varepsilon_{\mathrm{F}})\frac{\partial f_0}{\partial\varepsilon}\,\mathrm{d}^3\boldsymbol{k} \tag{10.12}$$

and similarly if we use (9.4) in (10.1), we find

$$\boldsymbol{Q}=(e/4\pi^3)\int \boldsymbol{v}_{\boldsymbol{k}}(\boldsymbol{L}_{\boldsymbol{k}}\cdot\boldsymbol{E})(\varepsilon_{\boldsymbol{k}}-\varepsilon_{\mathrm{F}})\frac{\partial f_0}{\partial\varepsilon}\,\mathrm{d}^3\boldsymbol{k} \tag{10.13}$$

Exactly as we did with σ and κ, we can now write $J_i=\sum_j C_{ij}G_j$, $Q_i=\sum_j D_{ij}E_j$, and consider the mean values C and D where $C=\frac{1}{3}(C_{xx}+C_{yy}+C_{zz})$, etc. We thus find

$$CT=(e/12\pi^3)\int \boldsymbol{v}_{\boldsymbol{k}}\cdot\boldsymbol{L}_{\boldsymbol{k}}^T(\varepsilon_{\boldsymbol{k}}-\varepsilon_{\mathrm{F}})\frac{\partial f_0}{\partial\varepsilon}\,\mathrm{d}^3\boldsymbol{k} \tag{10.14}$$

and

$$D=(e/12\pi^3)\int \boldsymbol{v}_{\boldsymbol{k}}\cdot\boldsymbol{L}_{\boldsymbol{k}}(\varepsilon_{\boldsymbol{k}}-\varepsilon_{\mathrm{F}})\frac{\partial f_0}{\partial\varepsilon}\,\mathrm{d}^3\boldsymbol{k} \tag{10.15}$$

From (10.14) and (10.15), we expect to have $CT=D$ whenever

$\boldsymbol{L}_k^T = \boldsymbol{L}_k$, that is, at high temperatures ($T \geqslant \theta_D$) and at very low temperatures, where defect scattering dominates. In fact it can be shown, surprisingly, that $CT = D$ at all temperatures, even when $\boldsymbol{L}_k^T \neq \boldsymbol{L}_k$: in both (10.14) and (10.15), the contributions from states above and below the FS almost exactly cancel, and it can be shown that the small residue is the same in both cases.

Replacing $\mathrm{d}^3\boldsymbol{k}$ by $\mathrm{d}\varepsilon\,\mathrm{d}S_k/\hbar v_k$ as usual, (10.15) becomes

$$CT = D = (e/12\pi^3\hbar)\int(\varepsilon_k - \varepsilon_F)\frac{\partial f_0}{\partial \varepsilon}\,\mathrm{d}\varepsilon\,\boldsymbol{L}_{k,\parallel}\,\mathrm{d}S_k$$

$$= \frac{1}{e}\int(\varepsilon - \varepsilon_F)\frac{\partial f_0}{\partial \varepsilon}\sigma(\varepsilon)\,\mathrm{d}\varepsilon \qquad (10.16)$$

with $\sigma(\varepsilon)$ given by (9.11). Now $\int_{-\infty}^{\infty}(\varepsilon - \varepsilon_F)(\partial f_0/\partial\varepsilon)\,\mathrm{d}\varepsilon = 0$, by symmetry, so that if we write $\sigma(\varepsilon) = \sigma(\varepsilon_F) + (\varepsilon - \varepsilon_F)(\mathrm{d}\sigma/\mathrm{d}\varepsilon)_{\varepsilon_F} + \cdots$, the $\sigma(\varepsilon_F)$ term contributes nothing, and we are left with

$$CT = D = \frac{1}{e}\int(\varepsilon - \varepsilon_F)^2\frac{\partial f_0}{\partial \varepsilon}\left[\frac{\mathrm{d}\sigma(\varepsilon)}{\mathrm{d}\varepsilon}\right]_{\varepsilon_F}\mathrm{d}\varepsilon$$

$$= -(\pi^2k^2T^2/3e)(\mathrm{d}\sigma(\varepsilon)/\mathrm{d}\varepsilon)_{\varepsilon_F} \qquad (10.17)$$

using (10.7). The sign of C and D thus depends on the sign of $\mathrm{d}\sigma/\mathrm{d}\varepsilon$ – that is, on the extent to which the current is carried more by hot electrons or by cool ones.

In general, we may have both an electric field and a temperature gradient present, and $f(\boldsymbol{k}) - f_0(\boldsymbol{k})$ will thus be the sum of two contributions: one due to $\boldsymbol{E}$ and given by (9.4), the other due to $\boldsymbol{G}$ and given by (10.3). It follows at once that the currents $\boldsymbol{J}$ and $\boldsymbol{Q}$ are then given by

$$\boldsymbol{J} = \sigma\boldsymbol{E} + C\boldsymbol{G} \qquad (10.18)$$

$$\boldsymbol{Q} = D\boldsymbol{E} + \kappa\boldsymbol{G} \qquad (10.19)$$

where we assume for simplicity that the metal is cubic. (Otherwise we shall have to write $J_i = \sum\sigma_{ij}E_j + \sum C_{ij}G_j$, etc.) In discussing the thermoelectric effects, it is convenient to re-express (10.18) and (10.19) in terms of $\boldsymbol{J}$ and $\boldsymbol{G}$ as independent variables; we then find (problem 10.2)

$$\boldsymbol{E} = \rho \boldsymbol{J} - S\boldsymbol{G} \tag{10.20}$$

$$\boldsymbol{Q} = \Pi \boldsymbol{J} + \kappa_c \boldsymbol{G} \tag{10.21}$$

where

$$\rho = 1/\sigma, \quad S = C/\sigma, \quad \Pi = D/\sigma \quad \text{and} \quad \kappa_c = \kappa - \sigma S \Pi$$

Thus if $\boldsymbol{J} = 0$, the measured thermal conductivity will be κ_c rather than κ (and we return to this point later), and the heat flow $\boldsymbol{Q}$ will be accompanied by an electric field $\boldsymbol{E} = -S\boldsymbol{G}$, where the *thermoelectric power* S is given by

$$S = -(\pi^2 k^2 T/3e\sigma)(\mathrm{d}\sigma(\varepsilon)/\mathrm{d}\varepsilon)_{\varepsilon_F} \tag{10.22}$$

An ordinary thermocouple (Fig. 10.2) consists of two wires of different materials A and B joined at each end, with the junctions at different temperatures T_H and T_L. A thermoelectric emf V_{12} appears across the break in wire A, at temperature T_R. If x denotes distance measured along the wire, we have

$$V = -\int E_x \,\mathrm{d}x = -\int E_x(\mathrm{d}x/\mathrm{d}T)\,\mathrm{d}T = -\int S\,\mathrm{d}T$$

so that

$$V_{12} = -\left[\int_{T_R}^{T_H} S_A\,\mathrm{d}T + \int_{T_H}^{T_L} S_B\,\mathrm{d}T + \int_{T_L}^{T_R} S_A\,\mathrm{d}T\right]$$
$$= \int_{T_L}^{T_H} (S_B - S_A)\,\mathrm{d}T \tag{10.23}$$

A closely related effect occurs if the circuit of Fig. 10.2 is all at the same temperature, and a current is driven round the circuit by

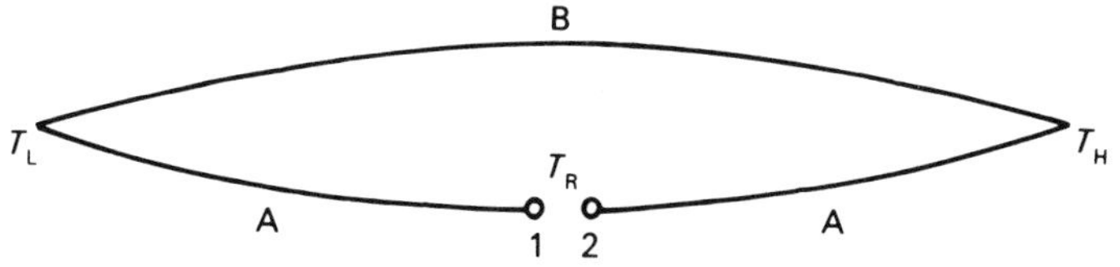

Fig. 10.2 Thermocouple, consisting of two different metals A and B.

applying a voltage across the terminals 1 and 2. Then $\boldsymbol{G}=0$ and (10.21) reduces to $\boldsymbol{Q}=\Pi \boldsymbol{J}$, where Π is the *Peltier coefficient*. Now if the cross-sectional area of the wire at any point is a, the total electric current will be $I=aJ$, and the total heat current will be $\Phi=aQ$, so that $\Phi=\Pi I$. The current I will be the same at all points around the circuit (even if a varies) so that if the whole circuit is made of the same material, with the same value of Π, Φ will also be the same at all points, and this uniform flow of heat energy will not be directly observable. But at a junction between two materials A and B, there will be a sudden change in Φ, and hence an evolution or absorption of heat at a rate

$$\Delta\Phi=(\Pi_{\mathrm{B}}-\Pi_{\mathrm{A}})I \tag{10.24}$$

This is called the Peltier effect.

The two quantities Π and S are not independent. It was shown thermodynamically by Lord Kelvin and (more rigorously) by Onsager that they must be related by the equation

$$\Pi=TS \tag{10.25}$$

Since $S=C/\sigma$ and $\Pi=D/\sigma$, this is precisely equivalent to the equation $CT=D$ that we discussed between (10.15) and (10.16); that equation is therefore a thermodynamic necessity.

From (10.23) and (10.24), we can find the difference $S_{\mathrm{B}}-S_{\mathrm{A}}$ by measuring $\mathrm{d}V_{12}/\mathrm{d}T_{\mathrm{H}}$, and the difference $\Pi_{\mathrm{B}}-\Pi_{\mathrm{A}}$ by measuring $\Delta\Phi/I$. We can find S for a single metal, instead of the difference between two metals, by measuring the *Thomson coefficient* μ. Thermodynamic arguments show that when both $\boldsymbol{J}$ and $\boldsymbol{G}$ are present simultaneously, there is a reversible absorption or evolution of heat in the metal, in addition to the irreversible Joule heating, at a rate $\mu\boldsymbol{J}\cdot\boldsymbol{G}$ per unit volume, where

$$\mu=T\,\mathrm{d}S/\mathrm{d}T \tag{10.26}$$

The same result follows if we use (10.20), (10.21) and (10.25) to work out the heat generated per unit volume, $\boldsymbol{J}\cdot\boldsymbol{E}-\operatorname{div}\boldsymbol{Q}$; we find

$$\boldsymbol{J}\cdot\boldsymbol{E}-\operatorname{div}\boldsymbol{Q}=\rho J^{2}+\mu\boldsymbol{J}\cdot\boldsymbol{G}-\operatorname{div}(\kappa_{\mathrm{c}}\boldsymbol{G}) \tag{10.27}$$

The quantity μ is a good deal more difficult to measure experiment-

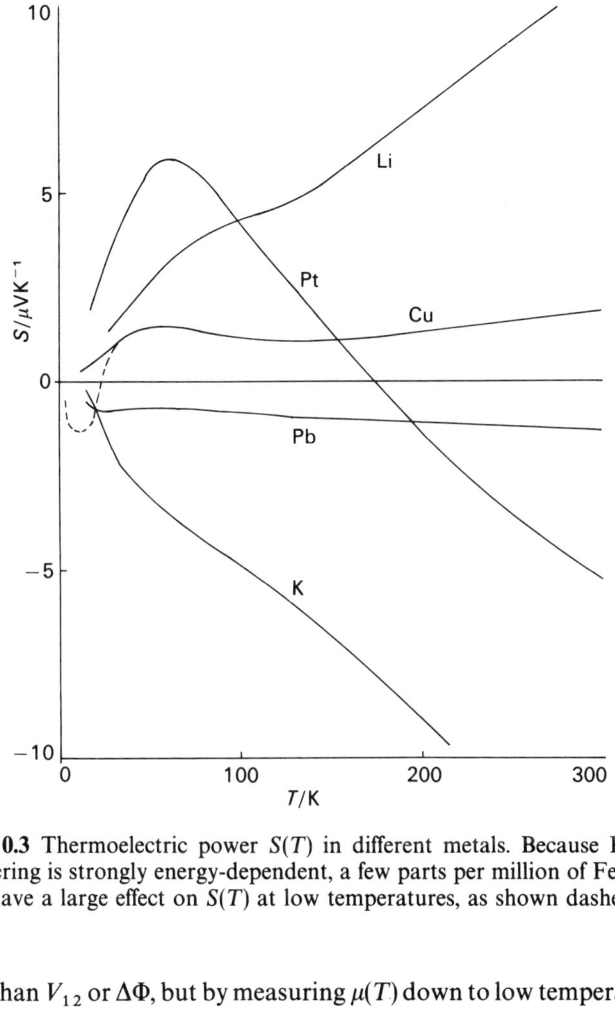

Fig. 10.3 Thermoelectric power $S(T)$ in different metals. Because Kondo scattering is strongly energy-dependent, a few parts per million of Fe in Cu can have a large effect on $S(T)$ at low temperatures, as shown dashed.

ally than V_{12} or $\Delta\Phi$, but by measuring $\mu(T)$ down to low temperatures on just one metal A and evaluating $\int(\mu/T)\,\mathrm{d}T$, we can find $S(T)$ for that metal, and then find $S(T)$ for any other metal B from the difference $S_B - S_A$. Usually Pb is used as the reference metal A.

The form of $S(T)$ varies a good deal from one metal to another, as shown in Fig. 10.3. Its magnitude is roughly what we might expect from (10.22): if we write $(\mathrm{d}\sigma(\varepsilon)/\mathrm{d}\varepsilon)_{\varepsilon_F} = \alpha\sigma(\varepsilon_F)/\varepsilon_F$, where α is a dimensionless parameter, (10.22) becomes

$$S = -\,\alpha(\pi^2 k/3e)(kT/\varepsilon_F) \tag{10.28}$$

and the observed values of S at 300 K correspond to values of α ranging from $+4$ for K to -7 for Li. It is not too surprising that $d\sigma/d\varepsilon$ and hence α should be positive for some metals and negative for others, as we can see from (9.11), which we can write as

$$\sigma(\varepsilon) = (e^2/12\pi^3\hbar)\bar{L}(\varepsilon)S_k(\varepsilon) \tag{10.29}$$

Here $\bar{L}(\varepsilon)$ is the average value of $L_{k,\parallel}$ over the constant-energy surface ε, and $S_k(\varepsilon)$ is the area of that surface in $\boldsymbol{k}$-space. Consider the contribution to $\sigma(\varepsilon)$ from just one energy band, in which ε ranges from ε_{min} to ε_{max}. For $\varepsilon < \varepsilon_{min}$, the band is completely empty and $S_k(\varepsilon) = 0$, and for $\varepsilon > \varepsilon_{max}$ it is completely full and again $S_k(\varepsilon) = 0$. As ε grows from ε_{min} to ε_{max}, $S_k(\varepsilon)$ starts from zero, passes through a maximum, and then falls again to zero. At the same time, the electron velocity $\boldsymbol{v}_k$ will be small for ε close to ε_{min}, growing as ε increases and then falling again as ε approaches ε_{max} (cf. Fig. 4.2), so that we can expect $L = \tau v$ to behave in the same way, though the precise variation will also depend on the variation of τ with ε. Thus both $\bar{L}(\varepsilon)$ and $S_k(\varepsilon)$ vary in a way that makes $d\sigma/d\varepsilon$ positive at some energies and negative at others. What is more surprising is that α differs in sign for two rather similar alkali metals, and that it should be so large in magnitude. From (10.29), we would normally expect $|\alpha| \gtrsim 2$ or so, as it is in Cu ($\alpha \approx -1.7$) and Pb ($\alpha \approx 1.7$).

From (10.28), we would expect $S \propto T$ (and hence $\mu \propto T$, from (10.26)), if α is temperature-independent. Figure 10.3 shows that this is so, roughly, at room temperature, but that at low temperatures a 'hump' appears in $S(T)$. This is due to 'phonon drag', an effect that we have so far neglected. When a current $\boldsymbol{J}$ is flowing through a metal, electron–phonon collisions will tend to create more phonons moving from left to right (say) than from right to left; in other words, to set up a heat current $\boldsymbol{Q}_\phi$ carried by the phonons, in addition to the heat current $\Pi\boldsymbol{J}$ carried by the electrons, which we now write as $\boldsymbol{Q}_e = \Pi_e\boldsymbol{J}$. If $\boldsymbol{Q}_\phi = \Pi_\phi\boldsymbol{J}$, the total heat current $\boldsymbol{Q} = \boldsymbol{Q}_e + \boldsymbol{Q}_\phi$ is given by $\boldsymbol{Q} = \Pi\boldsymbol{J}$, with $\Pi = \Pi_e + \Pi_\phi$, which may be very different from the electronic contribution Π_e alone. In the same way, a temperature gradient $\boldsymbol{G}$ will produce an imbalance in the phonon distribution, with more phonons flowing one way than the other, and this will lead, via electron–phonon collisions, to an imbalance in the electron distribution and hence to a phonon-induced current $\boldsymbol{J}_\phi = \sigma S_\phi \boldsymbol{G}$ in addition to the current $\boldsymbol{J}_e = \sigma S_e \boldsymbol{G}$ of (10.18). Again, we can still write $\boldsymbol{J} = \sigma S\boldsymbol{G}$, with $S = S_e + S_\phi$, and in fact the thermodynamic relation $\Pi = TS$

must still hold. The resultant phonon-drag contribution to S and Π falls off both at high temperatures and at very low temperatures: at low temperatures because the thermal energy carried by the phonons then becomes very small, and at high temperatures because collisions of phonons with other phonons are then frequent enough to make the phonon heat current $\boldsymbol{Q}_\phi$ very small.

As we saw in discussing (10.21), the thermal conductivity as normally measured will be $\kappa_c = \kappa - \sigma S\Pi$, rather than κ itself. Using $\Pi = TS$, we can write

$$\kappa_c = \kappa - \sigma S^2 T = \kappa(1 - S^2/\mathscr{L}) \tag{10.30}$$

(where $\mathscr{L} = \kappa/\sigma T$). Now normally $\mathscr{L}$ does not differ greatly from $\mathscr{L}_0 = \pi^2 k^2/3e^2$, so that from (10.28) we have $S^2/\mathscr{L} \approx \alpha^2(\pi^2/3)(kT/\varepsilon_F)^2$. In metals, the difference between κ and κ_c is thus negligibly small, but in semiconductors it can become quite large.

11

Magnetoresistance and Hall effect

11.1 BASIC IDEAS: THE FREE-ELECTRON MODEL

We saw in section 2.4 that a free-electron metal shows no magnetoresistance: ρ is independent of $\boldsymbol{B}$. All real metals, however, show a positive magnetoresistance: $\rho(\boldsymbol{B}) > \rho(0)$. The longitudinal magnetoresistance, $\rho_{\parallel}(\boldsymbol{B})$ (for $\boldsymbol{B} \parallel \boldsymbol{J}$) normally saturates at large $\boldsymbol{B}$; that is, it tends to a constant value not very much greater than $\rho(0)$. The behaviour of the transverse magnetoresistance, $\rho_{\perp}(\boldsymbol{B})$ (for $\boldsymbol{B} \perp \boldsymbol{J}$) is much more varied: in some metals it saturates, like $\rho_{\parallel}$; in others it grows indefinitely as B^2; in others, again, it seems to grow slowly and linearly with $|\boldsymbol{B}|$ at high fields, rather than saturating (Fig. 11.1). And in some metals the behaviour depends critically on the angle between $\boldsymbol{B}$ and the crystal axes: as $\boldsymbol{B}$ is rotated in the plane normal to $\boldsymbol{J}$, the behaviour switches abruptly back and forth between saturation and a quadratic rise (Fig. 11.2).

The Hall coefficient R_H, too, shows much more variety than the free-electron model would predict – it can be either positive or negative, and it can indeed change sign as $\boldsymbol{B}$ increases.

Almost all these effects can be understood in terms of the different kinds of orbit that can exist in a magnetic field; as we saw in section 7.2, these include closed electron orbits, closed hole orbits and open orbits. In this chapter, we shall try to show how they can account for the variety of effects observed.

To find the conductivity $\sigma(\boldsymbol{B})$ in a field $\boldsymbol{B}$, from which we can then find $\rho(\boldsymbol{B})$ and R_H, we proceed exactly as in section 9.1. We still have $f(\boldsymbol{k}) = f_0(\varepsilon - \Delta\varepsilon)$ (as in (9.5)), where $\Delta\varepsilon = -e\boldsymbol{E}\cdot\boldsymbol{L}_k$ is the energy picked up by the electron from the field $\boldsymbol{E}$ on its way to the point where the current is being calculated. It follows that the conductivity compo-

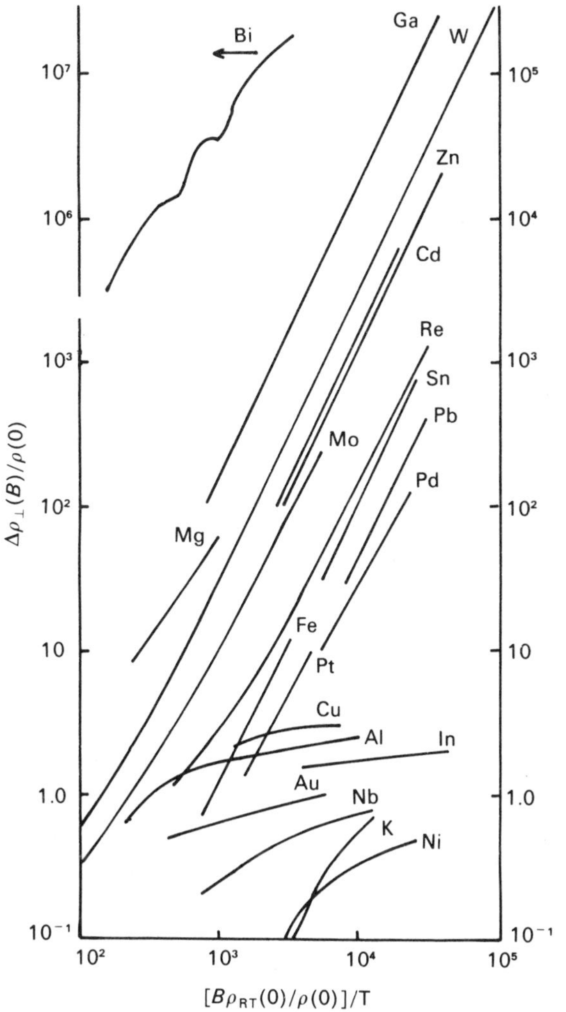

Fig. 11.1 Kohler plot showing $\Delta\rho_{\perp}(B)/\rho(0)$ against $B\rho_{RT}(0)/\rho(0)$ for different metals in transverse fields. ($\Delta\rho_{\perp}(B) = \rho_{\perp}(B) - \rho(0)$; $\rho_{RT}(0)$ is the value of $\rho(0)$ at room temperature). [From Fawcett (1964), *Adv. Phys.* **13**, 176.]

nents σ_{ij} are still given by (9.8), as before, but $\boldsymbol{L}_k$ for any state $\boldsymbol{k}$ now differs from its zero-field value, because of the curvature of the electron orbits by $\boldsymbol{B}$, and it is this change in $\boldsymbol{L}_k$ which causes the changes in σ_{ij}.

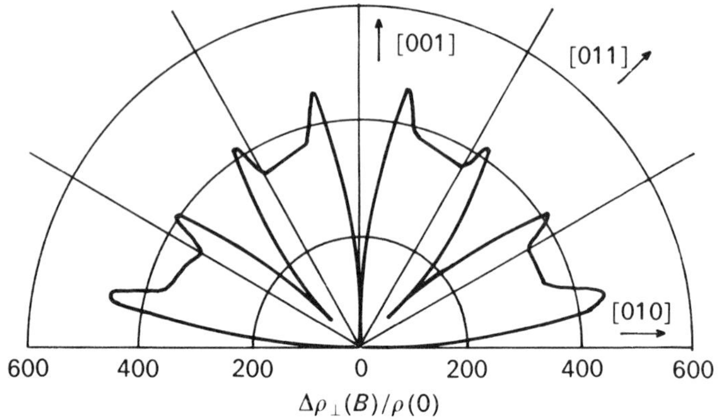

Fig. 11.2 Polar diagram showing $\Delta\rho_\perp(B)/\rho(0)$ for Cu crystal at 4.2 K (with $\rho_{RT}(0)/\rho(0) = 4000$) in a field $\boldsymbol{B}$ of 1.8 T, as $\boldsymbol{B}$ is rotated in the plane normal to $\boldsymbol{J}$. For most field directions, $\Delta\rho_\perp(B)/\rho(0)$ has risen quadratically to a very large value; for a few field directions, it has saturated at a much lower value. [From Klauder and Kunzler (1960), in *The Fermi Surface*, (ed. Harrison and Webb), Wiley, p. 125.]

To find σ_{ij}, we can use the expression

$$\sigma_{ij} = (e^2/4\pi^3\hbar)\int_{\mathrm{FS}} (v_i L_j/v)\mathrm{d}S_k \tag{11.1}$$

which follows from (9.8) in precisely the same way that (9.12) follows from (9.9). In a magnetic field B_z, it is convenient to re-express $\mathrm{d}S_k$ in terms of $\mathrm{d}k_z$, parallel to B_z, and $\mathrm{d}k_t$, measured around the orbit in the plane $k_z =$ constant. If $\boldsymbol{v}_n$ is the component of $\boldsymbol{v}$ normal to z, it follows

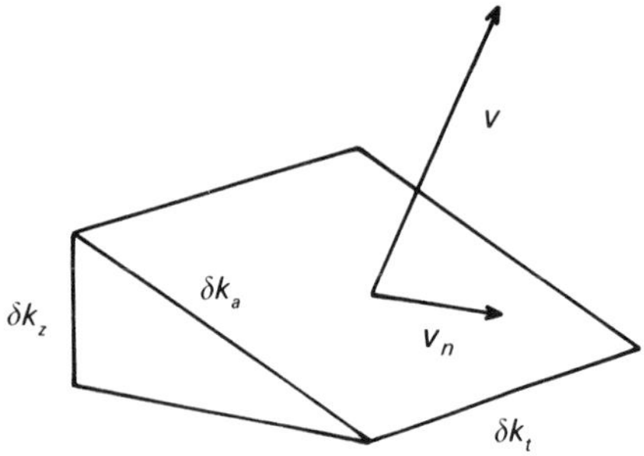

Fig. 11.3 The element of FS area is $\delta S_k = \delta k_t \delta k_a = \delta k_t \delta k_z v/v_n$.

from Fig. 11.3 that $\mathrm{d}S_k = \mathrm{d}k_z\,\mathrm{d}k_t v/v_n$, so that (11.1) becomes

$$\sigma_{ij} = (e^2/4\pi^3\hbar)\int \mathrm{d}k_z \oint \mathrm{d}k_t (v_i L_j/v_n) \tag{11.2}$$

where the integral $\oint \mathrm{d}k_t$ around the orbit (or orbits) in plane k_z is followed by integration over all orbits, i.e. all k_z.

Experimentally, we measure the resistivity components ρ_{ij} rather than the conductivity components σ_{ij}. If we assume that (for $\boldsymbol{B} = B_z$) σ_{xz}, σ_{yz}, σ_{zx} and σ_{zy} are all negligibly small (which should usually be a reasonably good approximation), inversion of the conductivity tensor leads to

$$\begin{aligned} &\rho_{xx} = \sigma_{yy}/\Delta, \quad \rho_{yy} = \sigma_{xx}/\Delta, \quad \rho_{yx} = -\sigma_{yx}/\Delta, \\ &\rho_{xy} = -\sigma_{xy}/\Delta, \quad \rho_{zz} = 1/\sigma_{zz} \end{aligned} \tag{11.3}$$

where $\Delta = \sigma_{xx}\sigma_{yy} - \sigma_{xy}\sigma_{yx}$, and all the other components vanish (problem 11.1). If we also assume that the x, y axes are chosen so that $\sigma_{xy} = \sigma_{yx} = 0$ when $\boldsymbol{B} = 0$ (and this can always be done), the components σ_{xy}, σ_{yx} and ρ_{xy}, ρ_{yx} represent the Hall effect. Symmetry then requires that $\sigma_{xy} = -\sigma_{yx}$, $\rho_{xy} = -\rho_{yx}$, so that the second term in Δ is always positive. The Hall coefficient is given by $R_\mathrm{H} = E_y/B_zJ_x = \rho_{yx}/B_z$ if $\boldsymbol{J} = J_x$, or (equally) by $R_\mathrm{H} = -E_x/B_zJ_y = -\rho_{xy}/B_z$ if $\boldsymbol{J} = J_y$.

We now need to see how $\boldsymbol{L}_k$ is affected by $\boldsymbol{B}$. Let us work out first, because it is rather easier to visualize, the average distance $\boldsymbol{L}^+$ that an electron will travel *after* time $t = 0$ before its motion is randomized. We assume for simplicity that every collision is fully randomizing, so that $\boldsymbol{L}^+$ is simply the average distance travelled after time $t = 0$ before the next collision, and we assume that the effective relaxation time τ is constant around the orbit. The electron then has a probability $\mathrm{e}^{-t/\tau}$ of surviving without collision at least until time t, at which time its velocity is $\boldsymbol{v}(t)$, so that its probable contribution to $\boldsymbol{L}^+$ between t and $t + \delta t$ is $\mathrm{e}^{-t/\tau}\boldsymbol{v}(t)\,\delta t$. We thus have the simple result

$$\boldsymbol{L}^+ = \int_0^\infty \mathrm{e}^{-t/\tau}\boldsymbol{v}(t)\,\mathrm{d}t \tag{11.4}$$

$\boldsymbol{L}$ is now given by a similar argument: for times $t < 0$, there is a probability $\mathrm{e}^{-|t|/\tau} = \mathrm{e}^{t/\tau}$ that the electron was already in the orbit at time t, so that

$$\boldsymbol{L} = \int_{-\infty}^{0} \mathrm{e}^{t/\tau} \boldsymbol{v}(t)\,\mathrm{d}t \tag{11.5}$$

If τ varies with $\boldsymbol{k}$ and hence varies around the orbit, the probability of survival from time t (< 0) until $t = 0$ will be given by $\exp - \int_t^0 \mathrm{d}u/\tau(u)$, where $1/\tau(u)$ is the effective scattering rate at the point $\boldsymbol{k}$ reached at time u, and the term $\mathrm{e}^{t/\tau}$ in (11.5) should then be replaced by this expression. For most purposes, though, we can ignore this complication, and assume τ constant around the orbit.

From the form of (11.2) and (11.5) we can deduce a useful general result: *Kohler's rule*, which says that for a given metal the ratio $\rho(\boldsymbol{B})/\rho(0)$ should depend only on the ratio $\boldsymbol{B}/\rho(0)$. (This means that magnetoresistance effects are best studied in pure metals at low temperatures; if we can reduce the zero-field resistivity $\rho(0)$ by a factor of ten, say, either by reducing the temperature or by using a purer sample, it will have the same effect on $\rho(\boldsymbol{B})/\rho(0)$ as increasing $\boldsymbol{B}$ by a factor of ten.) To derive Kohler's rule, we shall assume for brevity that τ is constant around each orbit (though it may vary from one orbit to the next), and that the orbits are closed. In fact the result still holds when τ varies around the orbit, and in the presence of open orbits, but is a bit more cumbersome to prove. We start from the fact that $\boldsymbol{v}(t)$ in (11.5) varies periodically as the electron moves around the orbit, so that we can write $\boldsymbol{v} = \boldsymbol{v}(\theta)$ where $\theta = 2\pi t/\tau_c = \omega_c t$: θ is thus a measure of position around the orbit. Writing $t/\tau = \theta/\omega_c\tau$, (11.5) then becomes $\boldsymbol{L} = \tau\int_{-\infty}^{0} \mathrm{e}^{\theta/\omega_c\tau}\boldsymbol{v}(\theta)\,\mathrm{d}\theta/\omega_c\tau$. Thus for a given orbit the ratio $\boldsymbol{L}/\boldsymbol{L}(0)$, where $\boldsymbol{L}(0) = \tau\boldsymbol{v}(0)$, depends on the field strength $\boldsymbol{B}$ and on τ only through the dimensionless product $\omega_c\tau$. Now $\omega_c\tau$ may vary from orbit to orbit, but for any orbit we have $\omega_c \propto B$, and if we also assume that for any orbit $\tau \propto 1/\rho(0)$, we have $\omega_c\tau \propto B/\rho(0)$. Since $\boldsymbol{L}$, integrated over the whole FS, determines $\sigma(\boldsymbol{B})$ and hence $\rho(\boldsymbol{B})$, and since $\boldsymbol{L}(0)$ likewise determines $\sigma(0)$ and $\rho(0)$, Kohler's rule follows.

In deriving it, though, we have assumed that τ varies in the same way with temperature for all orbits, and indeed for all states $\boldsymbol{k}$, so that we can write $\tau \propto 1/\rho(0)$. This is precisely the same assumption that underlies Matthiessen's rule, so that Kohler's rule and Matthiessen's rule have the same range of validity. The experimental results are consistent with this; Kohler's rule, like Matthiessen's rule, is a useful approximation, but is not obeyed exactly.

Now let us apply (11.5), first, to a free-electron metal. For free electrons in a field B_z, v_z is constant, so that (11.5) at once yields $L_z = v_z\tau$; the z component of $\boldsymbol{L}$ is unaffected by B_z, and so therefore are σ_{zz} and ρ_{zz}, from (11.2) and (11.3). In the plane normal to z, we can write

$$v_x(t) + iv_y(t) = v_0\,\mathrm{e}^{i(\omega_c t + \phi)} = (v_x + iv_y)\mathrm{e}^{i\omega_c t} \tag{11.6}$$

(where v_x, v_y are the values at $t = 0$), and then (11.5) gives

$$\begin{aligned} L_x + iL_y &= (v_x + iv_y)\int_{-\infty}^{0} \exp{(i\omega_c + 1/\tau)t}\,\mathrm{d}t \\ &= (v_x + iv_y)\tau/(1 + i\omega_c\tau) \end{aligned} \tag{11.7}$$

Using (11.2), we can thus write

$$\begin{aligned} \sigma_{xx} + i\sigma_{xy} &= (e^2/4\pi^3\hbar)\int \mathrm{d}k_z \oint \mathrm{d}k_t v_x(L_x + iL_y)/v_n \\ &= (e^2/4\pi^3\hbar)\int \mathrm{d}k_z \oint \mathrm{d}k_t v_x(v_x + iv_y)\tau/v_n(1 + i\omega_c\tau) \end{aligned} \tag{11.8}$$

The term in $v_x v_y$ vanishes by symmetry on integrating around the orbit, and we are left with

$$\sigma_{xx} + i\sigma_{xy} = \sigma_0/(1 + i\omega_c\tau) \tag{11.9}$$

where $\sigma_0 = \sigma_{xx}$ $(B = 0) = ne^2\tau/m$ (problem 11.2). This agrees completely with (2.20), since $\sigma_{xy} = -\sigma_{yx}$. It follows as in section 2.4 that $\rho_{xx} = \rho_0 = 1/\sigma_0$, and $R_H = -1/ne$, both independent of field strength.

As we saw in section 7.2, the essential difference between electron orbits and hole orbits is that the motion in $\boldsymbol{k}$-space and in real space is in opposite senses around the two: from (7.7), m_c^* and hence ω_c differ in sign for electron orbits and hole orbits, so that if a given field $\boldsymbol{B}$ produces anticlockwise motion around an electron orbit, it will produce clockwise motion around a hole orbit. The changed sign of ω_c means that in a metal containing only hole orbits, the Hall coefficient R_H will be *positive*. If we imagine for example a metal in which $\varepsilon = \varepsilon_0 + \hbar^2k^2/2m^*$, with m^* *negative*, so that the FS consists of a sphere enclosing n_h unfilled states (per unit volume of crystal), (11.9) still holds, with $\sigma_0 = n_h e^2\tau/|m^*|$ because we now have $\hbar k = |m^*|v$, but

$\omega_c = eB/m_c^*$ is now negative. When we invert (11.9) to find the resistivity tensor, we therefore find $\rho_{yx} = +B_z/n_h e$, so that

$$R_H = +1/n_h e \tag{11.10}$$

By showing how positive Hall coefficients can arise, this model resolves one of the major problems of the Sommerfeld free-electron model, but otherwise it is no more successful in explaining magnetoresistance – in both models ρ_{xx}, ρ_{zz} and R_H are all independent of field.

11.2 REAL METALS

To do better, we need a model in which the FS is something more realistic than just a single spherical surface. We consider in turn three models, which between them enable us to understand most of the observed behaviour: (i) a metal with a single, closed, non-spherical FS, holding n_e electrons or n_h 'holes' (unfilled states) per unit volume, (ii) a metal whose FS consists of a closed electron surface in one band, and a closed hole surface in another band, and (iii) a metal whose FS extends throughout $\boldsymbol{k}$-space in the periodic zone scheme, giving rise to the possibility of open orbits in the way discussed in section 7.2.

For the single, closed, non-spherical FS, consider first σ_{zz}, which is determined by an integral of $v_z L_z$ over the FS, as in (11.1) or (11.2). This integral can be re-expressed in terms of an average $\overline{v_z L_z}$ around each orbit. If $B_z = 0$, the electron does not move around the orbit between collisions, so that $L_z = v_z \tau$ and $\overline{v_z L_z} = \overline{v_z^2 \tau}$. But for $\omega_c \tau \gg 1$ (the 'high-field limit'), the electron will circulate many times around its orbit before colliding, during which time its average speed in the z direction is $\bar{v}_z$, so that $L_z \to \bar{v}_z \tau$, or $\bar{v}_z/\overline{(1/\tau)}$ if τ also varies around the orbit. In this limit, $\overline{v_z L_z}$ thus becomes $\bar{v}_z^2/\overline{(1/\tau)}$, which is always less than $\overline{v_z^2 \tau}$, though not by more than a factor of 2 or so unless v_z and τ vary a great deal around the orbit. We thus expect σ_{zz} for $\omega_c \tau \gg 1$ to be somewhat less than $\sigma_{zz}(0)$, and correspondingly ρ_{zz} for $\omega_c \tau \gg 1$ to be somewhat greater than $\rho_{zz}(0)$, but by no more than a factor of 2 or so, and this agrees well with experiment for all metals. Since the same behaviour will be expected for the other models that we shall look at, we do not need to look further at σ_{zz} and ρ_{zz}.

Now consider the transverse components σ_{xx}, σ_{yx}. For any closed orbit in $\boldsymbol{k}$-space, the corresponding real-space orbit in the xy plane will become smaller and smaller as B_z increases (cf. (7.9)), so that $\boldsymbol{L}_n$, the component of $\boldsymbol{L}$ in the xy plane, must likewise become smaller and smaller. From (11.5), we can write

$$\boldsymbol{L}_n = \int_{t=-\infty}^{0} \mathrm{e}^{-|t|/\tau} \mathrm{d}\boldsymbol{r}_n(t)$$

where $\mathrm{d}\boldsymbol{r}_n(t) = \boldsymbol{v}_n(t)\,\mathrm{d}t$, and we can think of $\boldsymbol{L}_n$ as the result of pursuing a path, backwards in time, around an exponentially shrinking version of the real-space orbit. Figure 11.4b shows the resultant spiral paths, for various $\omega_c\tau$, for an electron in the free-electron orbit shown in (a), and the dashed semicircle, which corresponds to (11.7), shows the resultant locus of $\boldsymbol{L}_n(\omega_c\tau)$. For $\omega_c\tau = 0$, $\boldsymbol{L}_n = \boldsymbol{L}_n(0) = \boldsymbol{v}_n(0)\tau$; as $\omega_c\tau$ increases, the angle between $\boldsymbol{L}_n$ and $\boldsymbol{L}_n(0)$ increases, and it is the component of $\boldsymbol{L}_n$ normal to $\boldsymbol{L}_n(0)$, $L_\perp$ say, which gives rise to the Hall effect. Figure 11.4d shows the corresponding spirals for the hole orbit shown in (c), which is the same as that shown in Fig. 4.5b, and again the dashed curve shows the locus of $\boldsymbol{L}_n(\omega_c\tau)$. We see that $L_\perp$ now changes sign for $\omega_c\tau \approx 0.3$; at lower fields the orbit 'looks like' an electron orbit, because the electron arriving at O has sampled only a small part of the orbit, and this small part curves in the same direction as an electron orbit. At larger $\omega_c\tau$, it has sampled more of the orbit, and $L_\perp$ changes to the expected sign. A star-shaped electron orbit, like that shown in Fig. 4.4 or 4.5c, can show the converse behaviour: an electron starting out in the re-entrant part of the orbit will initially 'see' it as a hole orbit, and only at higher fields as an electron orbit, and correspondingly the sign of $L_\perp$ will change from hole-like to electron-like as $\omega_c\tau$ increases.

We can thus see how σ_{xy} may change sign, and with it R_H, as the field increases. In fact the second-band hole surface in Al, shown in Fig. 5.4, will give rise to hole orbits rather like those of Fig. 11.4c, and indeed R_H in Al changes sign from negative to positive at $\omega_c\tau \approx 0.03$ (though in fact the main reason for that is somewhat different, as we shall see).

At high enough $\omega_c\tau$, R_H should always have the expected sign, and in fact it is possible to derive an explicit expression for R_H in the limit $\omega_c\tau \to \infty$. In this limit, the spirals of Fig. 11.4 will not diminish appreciably in size in one revolution, but they will eventually (as $t \to -\infty$) shrink to a common central point $\boldsymbol{r}_0$ say, wherever the electron is around the orbit at $t = 0$. It follows that an electron which is at point $\boldsymbol{r}_n$ at $t = 0$ will have travelled, on average, a distance $\boldsymbol{L}_n = \boldsymbol{r}_n - \boldsymbol{r}_0$, with

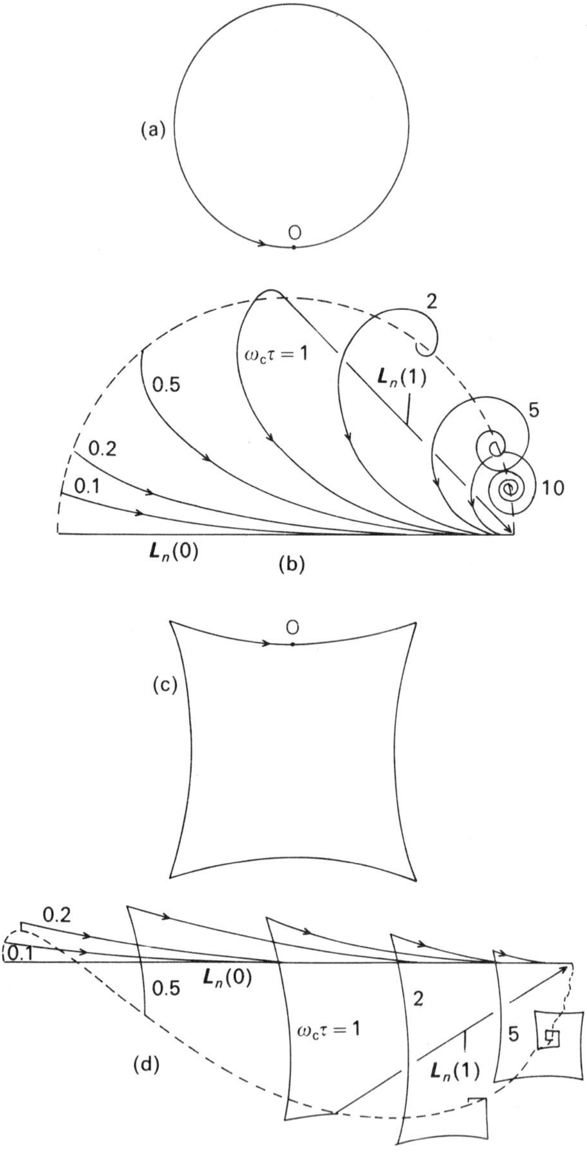

Fig. 11.4 An electron passing through point O on the real-space orbit (a) will effectively have pursued one of the spiral paths shown, for various $\omega_c\tau$, in (b). The resultant vector $\boldsymbol{L}_n(\omega_c\tau)$ joins the ends of the spiral, as shown for $\omega_c\tau = 1$. Similarly, an electron at O on the hole orbit (c) will effectively have pursued one of the paths shown in (d).

y component $L_y = y - y_0$. Using (7.9) we can write this as $L_y = -(\hbar/eB_z)(k_x - k_{x,0})$, where k_x refers to the corresponding point on the $\boldsymbol{k}$-space orbit, so that from (11.2),

$$\sigma_{xy} = -(e^2/4\pi^3\hbar)\int \mathrm{d}k_z \oint \mathrm{d}k_t(\hbar/eB_z)v_x(k_x - k_{x,0})/v_n$$

Now as the electron moves a distance δk_t around the $\boldsymbol{k}$-space orbit, a little thought shows that k_y changes by $\delta k_y = \pm v_x \delta k_t/v_n$ (+ for an electron orbit, − for a hole orbit), because the angle between v_x and v_n is the same as the angle between δk_y and δk_t. Thus

$$\begin{aligned}\sigma_{xy} &= \mp(e/4\pi^3 B_z)\int \mathrm{d}k_z \oint \mathrm{d}k_y(k_x - k_{x,0})\\ &= \mp(e/4\pi^3 B_z)\int \mathrm{d}k_z \mathscr{A}(k_z)\end{aligned}$$

where $\mathscr{A}(k_z)$ is the area of the $\boldsymbol{k}$-space orbit in the plane k_z,

$$= -n_e e/B_z \quad \text{or} \quad +n_h e/B_z \tag{11.11}$$

since $\int \mathrm{d}k_z \mathscr{A}(k_z)$ is simply the volume V_k enclosed by the FS, and $V_k/4\pi^3 = n_e$ for an electron surface, n_h for a hole surface.

Now if σ_{xx} and σ_{yy} vanish as B_z^{-2} at high fields, as we shall see they do in the absence of open orbits, it follows from (11.3) that $\rho_{yx} = 1/\sigma_{xy}$ in the high-field limit, so that

$$R_H = -1/n_e e \quad \text{or} \quad +1/n_h e \tag{11.12}$$

for an electron surface or a hole surface – a remarkably simple and general result.

We saw in section 7.2 that even if the FS is open, in the sense that it extends throughout $\boldsymbol{k}$-space in the periodic zone scheme, there will be many field directions which generate only closed electron or hole orbits, and no open orbits. For such field directions, the analysis leading to (11.11) and (11.12) is just as valid for open surfaces as for closed ones. It is also applicable to a metal containing several partially-filled bands, some containing electrons and some holes. The conductivity σ_{ij} is then the sum of contributions from each band, so that

$$\sigma_{xy} = (n_h - n_e)e/B_z \tag{11.13}$$

where n_h and n_e are the total numbers of holes and electrons, and (in the absence of open orbits)

$$R_H = 1/(n_h - n_e)e \tag{11.14}$$

No such simple and elegant result exists for the high-field limits of σ_{xx} and σ_{yy}. It is easy to see that they must vanish at least as fast as B_z^{-2} at high fields, if only closed orbits exist: $\boldsymbol{L}_n$ must get smaller and smaller as the orbit size diminishes, and so therefore must σ_{xx} and σ_{yy}, and since they must be even functions of $\boldsymbol{B}$, the result follows at once. It also follows at once from (11.3) that ρ_{xx} $(= \sigma_{yy}/\Delta)$ and ρ_{yy} $(= \sigma_{xx}/\Delta)$ must therefore saturate, since Δ $(\approx \sigma_{xy}^2)$ also vanishes as B_z^{-2}. As with ρ_{zz}, the saturation values depend on how far the FS departs from a free-electron sphere, and how much τ_k varies over the FS; detailed analysis shows that, as with ρ_{zz}, the saturation values are unlikely to exceed two or three times $\rho(0)$.

Experimentally, only a few metals show saturation of $\rho_\perp$ (i.e. ρ_{xx} or ρ_{yy}) for all field directions; these include the alkalis, Al and In. Even in these, saturation is often not complete: $\rho_\perp$ continues to rise slowly and more or less linearly with $|\boldsymbol{B}|$ at high fields. The rate of this linear rise varies from sample to sample, and it has long been a puzzle, but it is probably caused by sample imperfections (for example, a slightly wedge-shaped sample), which may effectively cause the measured $\rho_\perp$ to include a small spurious contribution $\alpha|\rho_H|$ from the 'Hall resistivity' ρ_H (i.e. ρ_{yx} or $-\rho_{xy}$). Typically, α needs to be only ≈ 0.01 to account for the apparent linear rise in $\rho_\perp$. (It is of course rather easy in experimental measurements to pick up a small contribution from ρ_H when measuring $\rho_\perp$ – it only needs a small misalignment of the potential probes applied to the sample – but what will then be measured is $\rho_\perp + \alpha\rho_H$, not $\rho_\perp + \alpha|\rho_H|$. The effect of such misalignment is easily eliminated by reversing the sign of $\boldsymbol{B}$, which reverses ρ_H, but that will not eliminate the linear term we are discussing here.)

Of the metals which do show saturation (apart from the residual slow linear rise), Al and In are more properly described by the *two-band* model: two closed surfaces, one containing electrons and the other holes. The properties of this model can easily be worked out, if we assume for simplicity that each band separately is isotropic in the xy plane. That is, we assume that the electron band, alone, would give rise to a resistivity which is the same in the x and y directions, $\rho_{xx} = \rho_{yy} = \rho_1$ say, and to a Hall resistivity $\rho_{yx} = -\rho_{xy} = R_1 B$ say; likewise the hole band, alone, would give rise to ρ_2 and $R_2 B$. (All these quantities, ρ_1, ρ_2, R_1 and R_2 may of course be field-dependent.) The conductivities of the two bands separately can then be compactly expressed in complex form: $\sigma_{xx,1} + i\sigma_{xy,1} = \sigma_{c,1} = 1/(\rho_1 - iR_1 B)$; $\sigma_{c,2} = 1/(\rho_2 - iR_2 B)$. The total conductivity of the metal is simply the sum of the conductivities of the two bands, $\sigma_{c,T} = \sigma_{c,1} + \sigma_{c,2}$, so that by writing $\sigma_{c,T} = 1/(\rho_T - iR_T B)$, we can express the resultant resistivity ρ_T and Hall coefficient R_T of the two-band model in terms of the contributions from the two bands separately. A little algebra (problem 11.3) shows that

$$\rho_T = \frac{\rho_1\rho_2(\rho_1 + \rho_2) + (\rho_1 R_2^2 + \rho_2 R_1^2)B^2}{(\rho_1 + \rho_2)^2 + (R_1 + R_2)^2 B^2} \tag{11.15}$$

$$R_T = \frac{\rho_1^2 R_2 + \rho_2^2 R_1 + R_1 R_2 (R_1 + R_2) B^2}{(\rho_1 + \rho_2)^2 + (R_1 + R_2)^2 B^2} \tag{11.16}$$

Thus in the two-band model, ρ_T and R_T show an explicit field-dependence, quite apart from any field-dependence of ρ_1, ρ_2, etc. themselves. The reason for this is clear enough, as we saw long ago (Fig. 2.6): if the currents $\boldsymbol{J}_1$ and $\boldsymbol{J}_2$ produced in the two bands by a field $\boldsymbol{E}$ flow in different directions, the resultant total current $|\boldsymbol{J}_T|$ will be less than $|\boldsymbol{J}_1| + |\boldsymbol{J}_2|$, and may be much less. Equations (11.15) and (11.16) merely represent the rather complicated outcome of adding $\boldsymbol{J}_1$ and $\boldsymbol{J}_2$ vectorially, and expressing $\boldsymbol{E}$ in terms of its components $E_\parallel = \rho_T J$ and $E_\perp = R_T B J$. If the two Hall angles ϕ_1 and ϕ_2 – the angles between $\boldsymbol{J}_1$ and $\boldsymbol{E}$ and between $\boldsymbol{J}_2$ and $\boldsymbol{E}$ – are equal, the explicit field-dependence vanishes. When $\phi_1 \neq \phi_2$, $\rho_T(B)$ will always be greater than $\rho_T(0)$ (problem 11.3). R_T will generally change sign as B increases, if R_1 and R_2 differ in sign, and this is in fact the main reason for the observed behaviour of Al.

At high fields, both ρ_T and R_T saturate, *unless* $R_1 + R_2 = 0$; in that case R_T becomes field-independent, but ρ_T grows without limit as B^2. To have $R_1 + R_2 = 0$ at high fields, (11.12) shows that we need a 'compensated' metal, in which $n_e = n_h$: the number of holes in band 2 must equal the number of electrons in band 1. Such compensation will occur naturally if the metal contains an even number of electrons per unit cell, because (4.9) shows that the electrons could then exactly fill an integral number of energy bands. In a metal, they partially fill two (or more) bands instead, and it is then clear that the total number of filled states in the upper band(s) must be exactly equal to the total number of empty states left behind in the lower band(s); in other words, $n_e = n_h$. This remains true even if there are more than two partly-filled bands (as in Fig. 4.5, for example), and even if the Fermi surfaces in some of the bands are open rather than closed. And if $n_e = n_h$ we still have $\rho_\perp \propto B^2$ (as long as $\boldsymbol{B}$ is in a direction that generates only closed orbits), even if n_e and n_h are divided among several partly-filled or partly-empty bands, basically because (11.13) shows that the B_z^{-1} term in σ_{xy} then vanishes. It follows that if σ_{xx}, $\sigma_{yy} \propto B_z^{-2}$, we must have $\Delta \propto B_z^{-4}$, and thus ρ_{xx}, $\rho_{yy} \propto B^2$.

If open orbits are present, we can no longer use this argument, because (11.11) and (11.13) are no longer valid. The derivation of these equations depends on the orbits being closed, so that they have a definite area $\mathscr{A}$, and breaks down in the presence of open orbits. We can therefore no longer argue that the B_z^{-1} term in σ_{xy} vanishes for a compensated metal, and in general it does not. But open orbits also have another and much more important effect: if there are open orbits running in say the k_x direction in $\boldsymbol{k}$-space, then the conductivity component σ_{yy} no longer falls as B_z^{-2} at high fields, but becomes independent of B_z. It is easy to see why this is. The $\boldsymbol{k}$-space orbit is still of limited width in the k_y direction, but it now extends indefinitely in the k_x direction, and correspondingly the real-space orbit is of limited width in the x direction – and this width gets smaller and smaller as B_z increases, from (7.9) – but it extends indefinitely in the y direction. It follows in the usual way that L_x and σ_{xx} fall to zero as B_z^{-2}, but L_y now behaves much more like L_z: the electron has a finite, non-vanishing average velocity $\bar{v}_y$ in the y direction, and $L_y \to \bar{v}_y \tau$ at high fields. This means that σ_{yy}, like σ_{zz}, tends to a finite limit at high fields, rather than vanishing as B_z^{-2}; from (11.3) it then follows that ρ_{xx} must grow as B_z^2, while ρ_{yy} saturates at a finite value. If the direction of the current $\boldsymbol{J}$ through the sample is at some angle θ to the x axis, the measured resistivity will be $\rho_{xx} \cos^2 \theta + \rho_{yy} \sin^2 \theta$, and will therefore grow as B_z^2 unless $\theta = \pi/2$. The open orbits act in effect as a one-dimensional conductor for current flow in the xy plane: they carry current readily in the y direction, but not in any other direction, so that the resistivity saturates in the y direction, but grows quadratically in every other direction.

We can now see how it is that in some metals the transverse resistivity $\rho_\perp$ varies from saturation to quadratic rise as $\boldsymbol{B}$ is rotated around $\boldsymbol{J}$; for some $\boldsymbol{B}$ directions open orbits are generated, and $\rho_\perp$ rises quadratically with $\boldsymbol{B}$ (unless $\boldsymbol{J}$ happens to be normal to the open-orbit axis in $\boldsymbol{k}$-space); for other directions only closed orbits exist, and $\rho_\perp$ saturates if the metal is uncompensated.

To summarize the results of this section, we show the field dependence of the transverse components of the conductivity and resistivity tensors,

$$\begin{vmatrix} \sigma_{xx} & \sigma_{xy} \\ \sigma_{yx} & \sigma_{yy} \end{vmatrix} \quad \text{and} \quad \begin{vmatrix} \rho_{xx} & \rho_{xy} \\ \rho_{yx} & \rho_{yy} \end{vmatrix}$$

(a) for an uncompensated metal ($n_e \neq n_h$) with no open orbits:

$$\begin{vmatrix} B^{-2} & B^{-1} \\ B^{-1} & B^{-2} \end{vmatrix} \quad \text{and} \quad \begin{vmatrix} B^0 & B \\ B & B^0 \end{vmatrix}$$

(b) for a compensated metal ($n_e = n_h$) with no open orbits:

$$\begin{vmatrix} B^{-2} & B^{-3} \\ B^{-3} & B^{-2} \end{vmatrix} \quad \text{and} \quad \begin{vmatrix} B^2 & B \\ B & B^2 \end{vmatrix}$$

(c) when there are open orbits running in the k_x direction:

$$\begin{vmatrix} B^{-2} & B^{-1} \\ B^{-1} & B^0 \end{vmatrix} \quad \text{and} \quad \begin{vmatrix} B^2 & B \\ B & B^0 \end{vmatrix}$$

Metals with closed Fermi surfaces can show only type (a) or (b) behaviour; metals containing open surfaces may switch from (a) to (c), or from (b) or (c), as the field direction is rotated. Some examples of metals showing each type of behaviour are:

(a) Li, K, Na (single-band metals); Al, In.
(b) Bi, As, Sb (semi-metals); Be, Mo, W.
(a) + (c) Cu, Ag, Au (single-band metals: cf. Fig. 5.7).
(b) + (c) Sn, Pb, Mg, Zn, Cd.

Finally, we should mention the phenomenon of 'magnetic breakdown', which complicates matters yet further. At high fields

B, an electron may be able to 'tunnel through' from one orbit in ***k***-space to another, if the two orbits come sufficiently close together. If it can switch orbits in this way, a whole new set of possibilities arises; for example, open orbits may become closed, and closed orbits may become open. This has all sorts of consequences both for magnetoresistance and for the de Haas–van Alphen effect, but it would take us too far afield to explore them here.

12

Radio-frequency, optical and other properties

12.1 RADIO-FREQUENCY PROPERTIES

We discussed the rf and optical properties of a free-electron metal in section 2.5, in terms of the conductivity σ_ω (eqn (2.27)) and the skin depth δ. At room temperature and above, the properties of real metals are described by essentially the same equations, suitably generalized: in (2.27), for example, σ_0 will then be given by (9.13), and τ must be replaced by some suitable average if τ_k varies with $\boldsymbol{k}$. But in pure metals at low temperatures and high frequencies, the mean free path of the electrons may become large compared with δ, and the treatment of section 2.5 then breaks down. In this section, we look at the behaviour of metals under these 'anomalous skin effect' conditions. First, though, we must see how to calculate the current density $\boldsymbol{J}$ produced by a field $\boldsymbol{E}$ when the field is non-uniform.

So far, we have expressed $\boldsymbol{J}$ (and hence σ) in terms of the vector mean free path $\boldsymbol{L}$; we have found the mean energy gained by an electron from the field $\boldsymbol{E}$ by writing

$$\Delta\varepsilon_k = -e\boldsymbol{E}\cdot\boldsymbol{L}_k \tag{12.1}$$

and then found $f(\boldsymbol{k})$ and hence $\boldsymbol{J}$ by writing $f(\boldsymbol{k}) = f_0(\varepsilon_k - \Delta\varepsilon_k)$ (eqn (9.5)). But when the field $\boldsymbol{E}(\boldsymbol{r}, t)$ varies appreciably over a free path, or over the time taken to traverse a free path, (12.1) is no longer valid. To simplify matters, we again assume (as in section 11.1) that every collision is fully randomizing, and we write $l_k = v_k\tau_k$ for the mean free path of an electron, in $\boldsymbol{B} = 0$, in state $\boldsymbol{k}$. Then it is fairly obvious how we need to generalize (12.1) in order to find $\Delta\varepsilon_k$ when the field $\boldsymbol{E}$ varies in space or time. If the electron is at point

$\boldsymbol{r}$ at time t, the rate at which it gains energy from the field is $-e\boldsymbol{E}(\boldsymbol{r},t)\cdot\boldsymbol{v}_k$, and the probability that it will then survive until $t=0$ without collision is e^{t/τ_k}. Thus

$$\Delta\varepsilon_k = -e\int_{-\infty}^{0} \boldsymbol{E}(\boldsymbol{r},t)\cdot\boldsymbol{v}_k\, e^{t/\tau_k}\,dt \tag{12.2}$$

which reduces at once to (12.1) if $\boldsymbol{E}$ is steady and uniform. In writing down (12.2), we have assumed that $\boldsymbol{B}=0$, so that $\boldsymbol{k}, \boldsymbol{v}_k$ and τ_k do not change with time. If $\boldsymbol{B}\neq 0$, we have to replace $\boldsymbol{v}_k e^{t/\tau_k}$ in (12.2) by $\boldsymbol{v}(t)\exp - \int_t^0 du/\tau(u)$, as in (11.5).

Equation (12.2) thus provides the generalization of (12.1) needed to discuss the anomalous skin effect and other problems involving non-uniform fields. In fact, the resulting equations for the anomalous skin effect are somewhat formidable, and we shall instead adopt an approximate approach which enables us to avoid them.

When an rf field $E_x(0)e^{i\omega t}$ is applied to the surface $z=0$ of a metal, it induces screening currents in the surface layers of the metal. If the current density at depth z is $J_x(z)\,e^{i\omega t}$, the response of the metal to the applied field is described by the *surface impedance*

$$Z = E_x(0)\bigg/\int_0^{\infty} J_x(z)dz \tag{12.3}$$

which is the ratio of the field at the surface to the total current, per unit width of metal, below the surface (so that Z has the dimensions of resistance). In general, the total current will not be in phase with the applied field, so that Z will be complex: $Z = R + iX$.

We can derive an alternative expression for Z by combining the Maxwell equations curl $\boldsymbol{E} = -\dot{\boldsymbol{B}}$ and curl $\boldsymbol{B} = \mu_0\boldsymbol{J}$ (neglecting the displacement current term $\dot{\boldsymbol{D}}$, as we almost always can in a metal, except at optical frequencies) to yield the expression

$$i\omega\mu_0 J_x = d^2E_x/dz^2 \tag{12.4}$$

from which it follows that $\int_0^\infty J_x(z)dz = -[dE_x/dz]_{z=0}/i\omega\mu_0$, so that

$$Z = -i\omega\mu_0 E_x(0)/[dE_x/dz]_0. \tag{12.5}$$

Under the 'classical' skin effect conditions assumed in section 2.5, the current density $J_x(z)$ at depth z is related to $E_x(z)$ by Ohm's law, and $E_x(z)$ falls off exponentially with z, so that

$$J_x(z) = \sigma_0 E_x(z) = \sigma_0 E_x(0) \exp - (1+i)z/\delta \tag{12.6}$$

where δ is the classical skin depth (2.33). From (12.3) or (12.5), we then have

$$Z = (1+i)/\sigma_0 \delta \tag{12.7}$$

But under anomalous skin effect conditions, Ohm's law no longer holds, and it follows from (12.2) that $J_x(z)$ at a particular depth z will be given by an integral involving $E_x(z)$ over a range of depths. Combining this with (12.4) yields a rather formidable-looking equation for $E_x(z)$, whose solution is a rather lengthy process. We can avoid this, and arrive at essentially the correct result for Z, by using an approximate argument based on Pippard's 'ineffectiveness concept'.

On this approach, we simplify the problem by assuming that the field $E_x(z)$ still falls off exponentially, with an effective skin depth δ_{eff}, and that $J_x(z)$ and $E_x(z)$ are still related by Ohm's law, with an effective conductivity σ_{eff}. Consequently the skin depth and the surface impedance are still given by (2.33) and (12.7), but with σ_0 and δ replaced by σ_{eff} and δ_{eff}. But in the 'anomalous limit' ($\delta_{\text{eff}} \ll l$), σ_{eff} is much smaller than the dc conductivity σ_0, because most electrons spend only a tiny fraction of a mean free path within the skin depth, and therefore pick up a negligible amount of energy from the rf field. Only electrons travelling almost parallel to the surface of the metal will spend an appreciable fraction of a mean free path exposed to the rf field, and we therefore assume that only these electrons, travelling within a range of angles $\pm \beta\delta_{\text{eff}}/l$ say (where $\beta \sim 1$) contribute to σ_{eff}.

The effect of this assumption is to make σ_{eff} independent of l: as l increases, the number of effective electrons falls as $1/l$, and this just cancels out the increase of σ with l that would otherwise occur. For free electrons, it is not difficult to show that $\sigma_{\text{eff}} = (3\beta\delta_{\text{eff}}/2)(\sigma_0/l)$, independent of l since $\sigma_0 \propto l$. More generally, we can find σ_{eff} from the expression

$$\sigma_{xx} = (e^2/4\pi^3\hbar) \int (v_x^2/v)_{\mathbf{k}} \tau_{\mathbf{k}} \, \mathrm{d}S_{\mathbf{k}}$$

$$= (e^2/4\pi^3\hbar) \int (v_x^2/v^2)_{\mathbf{k}} l_{\mathbf{k}} \, \mathrm{d}S_{\mathbf{k}} \tag{12.8}$$

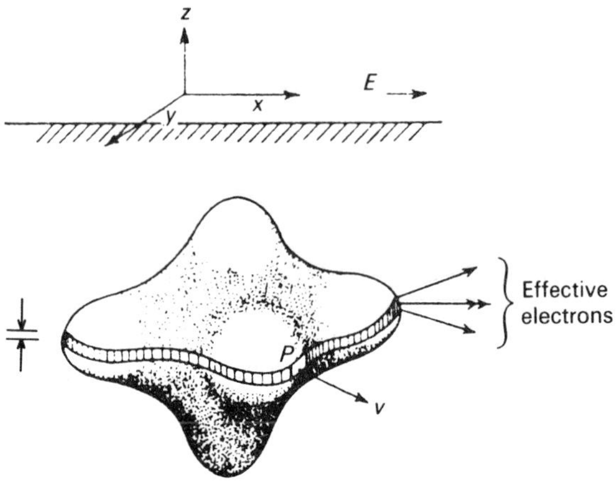

Fig. 12.1 Showing the narrow band of effective electrons, travelling parallel to the sample surface, in the anomalous skin effect. [From Ziman (1972), *Principles of the Theory of Solids*, 2nd edn, Cambridge University Press, London.]

which (like (11.1)) follows from (9.8) in the same way that (9.12) follows from (9.9). Integrated over the whole FS, (12.8) yields σ_{xx} in the normal way. To find σ_{eff}, we integrate only over that part of the FS on which the electrons are 'effective'; that is, travelling within the range of angles $\pm\beta\delta_{\text{eff}}/l$ to the surface of the metal.

Since $\boldsymbol{v}_k$ is normal to the FS at point $\boldsymbol{k}$, this means that the integration is confined to a narrow band around the 'equator' of the FS, as shown in Fig. 12.1. The width Δk_z of the effective band at any point $\boldsymbol{k}$ on the equator will depend on the curvature of the FS at that point. If the radius of curvature of the FS, in a plane containing $\boldsymbol{v}_k$ and z, is ρ_k say, then $\Delta k_z = 2\beta\delta_{\text{eff}}\rho_k/l_k$. Thus if k_t is measured around the equator, in a plane normal to z, we have $\mathrm{d}S_k = 2\beta\delta_{\text{eff}}\rho_k\,\mathrm{d}k_t/l_k$, so that

$$\sigma_{\text{eff}} = (e^2/4\pi^3\hbar)2\beta\delta_{\text{eff}}\oint\rho_k\cos^2\theta\,\mathrm{d}k_t, = 2\beta\delta_{\text{eff}}s \text{ say,} \qquad (12.9)$$

where θ is the angle between $\boldsymbol{v}_k$ and the x axis. As before, σ_{eff} is independent of the mean free path l_k, and we see that it depends only on δ_{eff} and on the geometry of the FS around the equator.

Knowing σ_{eff}, we can find δ_{eff} and Z on the ineffectiveness concept approach from (2.33) and (12.7), with σ_0 and δ replaced by σ_{eff} and

δ_{eff}. We thus find

$$Z = \tfrac{1}{2}(1+i)(\mu_0\omega)^{2/3}/(\beta s)^{1/3} \tag{12.10}$$

so that Z, like σ_{eff}, depends only on the geometry of the FS.

The expression (12.10) is identical in form with the result of an exact treatment, and can be made completely identical by appropriate choice of the parameter β. In the exact treatment, Z depends slightly on what is assumed to happen to the electrons when they hit the metal surface at $z = 0$ – whether they are scattered randomly or reflected specularly, as if from a mirror. If scattering is random, $\beta = 1.21(1-i)$, and if scattering is specular, $\beta = 1.72(1-i)$. It is not surprising that β turns out to be complex; from (12.9), this merely means that in our approximate treatment there is a phase difference between $J(z)$ and $E(z)$.

In pure metals at low temperatures, where l may be 0.1 mm or more, the anomalous limit $l \gg \delta_{eff}$ may be reached at frequencies as low as 1 MHz or less. In this limit, δ_{eff} will typically be about 5 μm at 1 MHz, falling to about 0.1 μm at microwave frequencies (problem 12.1).

The fact that Z is determined entirely by the quantity s means that a good deal can be learnt about the geometry of the FS by measuring Z (or its real part R, which determines the Q of a resonant cavity) on a number of single-crystal sample surfaces, cut at different angles to the crystal axes and so having different values of s. It was in this way that Pippard first established the shape of the FS of Cu (Fig. 5.7) in 1957. But this technique is a good deal less powerful than the de Haas–van Alphen effect, and has been little used since then.

Quite different information about the FS can be gained by measuring Z or R with a steady magnetic field $\boldsymbol{B}$ applied parallel to the metal surface. As B is varied, Z too varies, periodically in $1/B$, and this effect is called Azbel'–Kaner cyclotron resonance. The origin of the oscillations is clear from Fig. 12.2. The effect of $\boldsymbol{B}$ is to curl up the electron orbits into helical paths, so that they re-enter the skin depth at intervals $\tau_c = 2\pi/\omega_c$ (with ω_c given by (7.6)) until they suffer a collision. If $\omega_c\tau \gg 1$, they will complete many orbits before collision. Now whenever $\omega_c = \omega/n$, where ω is the frequency of the rf field applied to the surface, the electrons will see the field in the same phase at each traverse of the skin depth, and will therefore contribute particularly strongly to the rf current flowing in the skin

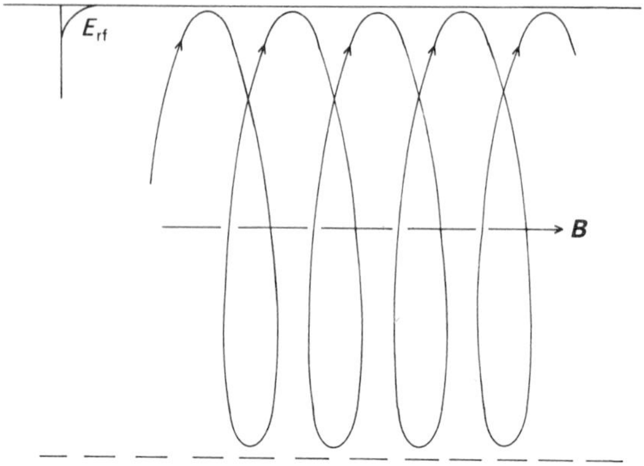

Fig. 12.2 Showing the origin of cyclotron resonance. The rf electric field penetrates below the surface only to a depth δ_{eff}, much less than the orbit radius. To observe the rf size effect a thin plate is used, as indicated by the dashed line, so that the orbit just fits between the two surfaces. [From Chambers (1956), *Can J. Phys.*, **34**, 1395.]

depth. We therefore expect to see oscillations in Z whenever $\omega_c = \omega/n$, i.e. whenever $1/B = en/m_c^*\omega$, precisely as observed. From the period $\Delta(1/B) = e/m_c^*\omega$, we can deduce the effective mass m_c^* of the electrons contributing to the signal. As we saw in section 7.2, different orbits, on different slice-planes k_B, will generally have different values of m_c^*, but there will be some values of k_B for which m_c^* goes through an extremum, and what will be observed experimentally will be these extremal values of m_c^*, where many orbits contribute simultaneously.

Cyclotron resonance experiments are normally carried out at low temperatures and at microwave frequencies, to make $\omega\tau$ as large as possible. If instead we measure Z at a much lower frequency, 10 MHz or so, on a thin parallel-sided plate of metal, with $\boldsymbol{B}$ parallel to the surface, we shall be doing a quite different experiment – we shall be observing the rf size effect, or Gantmakher effect. The cyclotron frequency will now be very much larger than ω, so that the electrons return to the skin depth many times in each rf cycle; as in cyclotron resonance, their contribution to the surface current is therefore enhanced. (In fact we can regard this as 'cyclotron resonance' with $n = 0$.) But this will only happen as long as the helical orbits can fit into the plate thickness d. If the 'caliper diameter' Δz of the orbit (the distance between the highest and lowest points on the orbit) is greater than d, the electrons will collide with one of the surfaces on their way round, and the enhancement factor is lost. The contribution to Z from electrons on a given orbit will therefore change abruptly, as B is varied, when $\Delta z = d$. Now in a given field B, different orbits, on different slice-planes k_B, will have different orbit diameters Δz and, just as in cyclotron resonance, the observed effects will come from the regions where Δz is extremal, so that many orbits contribute simultaneously. If $\boldsymbol{B} = B_y$ say, (7.9) shows that

$$\Delta z = (\hbar/eB_y)\Delta k_x \qquad (12.11)$$

where Δk_x is the $\boldsymbol{k}$-space caliper diameter of the orbit. By measuring the fields B_y at which Z changes abruptly, the Gantmakher effect thus enables us to measure the extremal caliper diameters Δk_x of the FS, and thus provides remarkably direct evidence on the FS geometry. But to use it successfully, we need to be able to grow thin single-crystal plates of very high purity, and to work at low temperatures, because the mean free path must be long enough to satisfy the condition $l \gg \delta_{\text{eff}}$ at the relatively low frequency of 10 MHz or so.

Clearly, both cyclotron resonance and the Gantmakher effect will become less and less observable as $\omega_c\tau$ falls, and the electrons complete fewer orbits before colliding. Indeed, the temperature-dependence of the effects can be used to determine the temperature-dependence of τ. When this is done, it is found that $1/\tau$ varies as $a + bT^3$, even when the dc resistivity is varying as $a + bT^5$, and the reason for this is not hard to see. In section 9.2 we saw that (under certain assumptions) phonon scattering produced a T^5 contribution to the effective scattering rate at low temperatures, of which a factor T^2 arose because the scattering was through only small angles, which had little effect in limiting the electric current. But in both cyclotron resonance and the Gantmakher effect, the electron needs to stay in a rather accurately defined orbit if it is to pass repeatedly through the skin depth, and even small-angle scattering will suffice to remove an electron from that orbit. The effective scattering rate therefore varies as T^3 rather than T^5, just as it does in thermal conductivity (section 10.1), and for similar reasons.

12.2 OPTICAL AND OTHER PROPERTIES

Up to infra-red frequencies, the optical properties of metals are usually well described by the simple Drude theory of section 2.5.

The theory can be improved in obvious ways to take account of the properties of real metals; for example, when $\omega\tau \gg 1$, so that (2.27) reduces to $\sigma_\omega = -i\sigma_0/\omega\tau = -ine^2/m\omega$, it is not difficult to show that for a real metal

$$\sigma_\omega = -i(e^2/12\pi^3\hbar\omega)\int_{FS} v_k \, dS_k \qquad (12.12)$$

(problem 12.2), as we might expect from (9.13). But these improvements do not alter the qualitative behaviour; in particular, the prediction that the real part of σ_ω should fall to zero as $1/\omega^2$ at high frequencies. Since $Re(\sigma_\omega)$ is responsible for the absorption of energy from the incident radiation, the energy absorption should also fall towards zero as the frequency increases. Experimentally, this is indeed what happens, until a certain frequency is reached, but then $Re(\sigma_\omega)$ starts to increase rapidly again, in a way that cannot be accounted for on the free-electron model. Figure 12.3 shows the corresponding absorption edge in $Im(\varepsilon_{\text{eff}})$, which is related to $Re(\sigma_\omega)$ by (2.31).

This absorption is caused by inter-band transitions, which we mentioned briefly in section 2.5. These inter-band transitions are themselves caused by a kind of internal photoelectric effect, in which a photon of the incident radiation gives up its energy to one of the conduction electrons and in doing so raises it from one energy band into another higher band. If the electron starts in state $\boldsymbol{k}$ in band m, say, with energy $\varepsilon_m(\boldsymbol{k})$, and ends in state $\boldsymbol{k}'$ in band n with energy $\varepsilon_n(\boldsymbol{k}')$, then we must have

$$\varepsilon_n(\boldsymbol{k}') = \varepsilon_m(\boldsymbol{k}) + \hbar\omega \quad \text{and} \quad \boldsymbol{k}' = \boldsymbol{k} + \boldsymbol{k}_\phi \qquad (12.13)$$

where $\hbar\omega$ is the energy of the photon and $\boldsymbol{k}_\phi$ its wave-vector. The first of these conditions is of course simply energy conservation, and the second is 'momentum' conservation of the kind we met in (8.14). But for visible light, or even for UV radiation, $\boldsymbol{k}_\phi$ is negligibly small compared with typical values of $\boldsymbol{k}, \boldsymbol{k}'$, so that momentum conservation reduces to $\boldsymbol{k}' = \boldsymbol{k}$. (Because $\boldsymbol{k}_\phi$ is so small, there is no need in the present problem to introduce the reciprocal lattice vector $\boldsymbol{K}$ that appeared in (8.14); here, $\boldsymbol{K}$ is always zero.)

In terms of an $\varepsilon(\boldsymbol{k})$ diagram such as Fig. 5.5 or 5.6, therefore, we have a vertical transition (since $\boldsymbol{k}' = \boldsymbol{k}$) from a filled state below ε_F to an empty state in a higher band, above ε_F. To produce such a

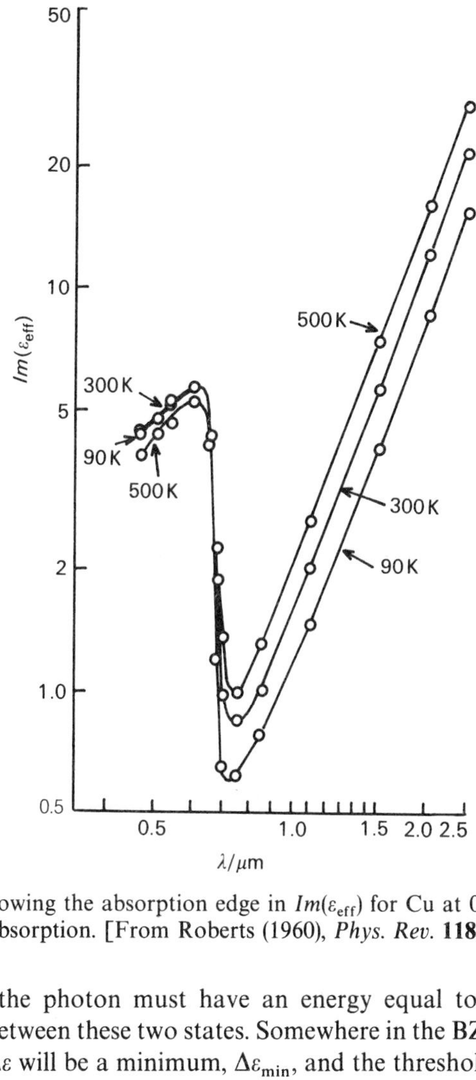

Fig. 12.3 Showing the absorption edge in $Im(\varepsilon_{eff})$ for Cu at 0.6 μm, due to inter-band absorption. [From Roberts (1960), *Phys. Rev.* **118**, 1509.]

transition, the photon must have an energy equal to the energy difference between these two states. Somewhere in the BZ this energy difference $\Delta\varepsilon$ will be a minimum, $\Delta\varepsilon_{min}$, and the threshold frequency at which inter-band absorption begins will be given by $\hbar\omega = \Delta\varepsilon_{min}$. In most metals, this frequency is at or beyond the blue end of the visible spectrum, but in Cu, for example, it is towards the red end of the spectrum, and it is the absorption of higher frequencies that gives Cu its characteristic colour. The value of $\Delta\varepsilon_{min}$ in Cu is low (about 2 eV) because the filled d bands lie only this distance below the FS; at ω_{min}, electrons begin to be excited from these states to states just above the FS, and this is what causes the sharp rise in $Im(\varepsilon_{eff})$ at wavelengths less than about 0.6 μm, as shown in Fig. 12.3.

If the incident radiation has a bandwidth $\delta\omega$, and some simplifying assumptions are made, the rate of energy absorption for $\omega > \omega_{min}$ should be proportional to $g(\omega)\delta\omega$, where $g(\omega)$ is the 'joint density of states' – the number of filled states, per unit energy range, which lie at distance $\hbar\omega$ below empty states of the same $\boldsymbol{k}$-value. For given ω, these states will lie on a surface S_ω in $\boldsymbol{k}$-space, and in fact $g(\omega)$ can be written in a form similar to (7.2):

$$g(\omega) = \int \mathrm{d}S_\omega/(4\pi^3\hbar|v_{n,\boldsymbol{k}} - v_{m,\boldsymbol{k}}|) \tag{12.14}$$

Thus measurements of the optical properties can yield, in principle, some information about the form of $\varepsilon(\boldsymbol{k})$ both above and below the FS, but in practice this information is not at all easy to decipher.

Considerably more direct information on $\varepsilon(\boldsymbol{k})$ can be obtained by looking at the photo-electrons themselves, if they have sufficient energy to escape through the metal surface into the world outside, and in recent years 'angle-resolved photo-emission spectroscopy' (ARPES) has become an important tool in the study of both the band structures and the surface properties of metals and semi-conductors.

Although the detailed theory of photo-emission is very complex, the basic physics can be quite simply understood. It requires a certain energy ε_w (called the 'work function' and typically 2–4 eV) to extract an electron at the Fermi surface from within the metal and place it, at rest, at a point outside. Its energy will then be $\varepsilon_F + \varepsilon_w = \varepsilon_s$ say. Thus if the energy $\varepsilon_n(\boldsymbol{k})$ of the electron, after absorbing the photon energy $\hbar\omega$, satisfies $\varepsilon_n(\boldsymbol{k}) \geqslant \varepsilon_s$ it will be able to escape from the metal (Fig. 12.4). We can measure ε_w by finding the threshold frequency ω_0 at which electrons first escape from the metal, as Millikan first did in 1916; we then have $\varepsilon_w = \hbar\omega_0$. Even for $\omega > \omega_0$, an electron will only escape, of course, if it is travelling in the right direction, and if it starts out not too far from the surface, because otherwise it will suffer a collision on the way. In fact an electron with energy 10–100 eV above ε_F – the kind of energy commonly used in ARPES experiments – will travel only a few nm before colliding, so that the photon absorption which produces it must occur very close to the surface. (This short mean free path is due to *electron–electron* collisions; for electrons close to the FS, these are normally not very

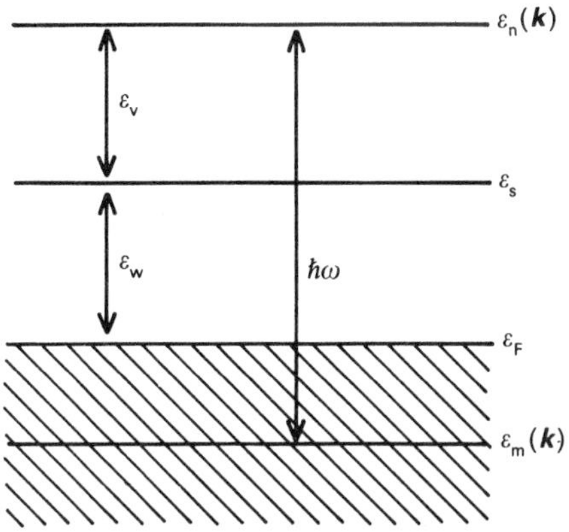

Fig. 12.4 The energy levels involved in photoemission.

important, and we have previously neglected them, but for an electron far from the FS they become increasingly important.)

If the electron does escape from the metal, its kinetic energy ε_v and direction of motion can be measured in the vacuum outside, and from these we can find its wave-vector, $\boldsymbol{k}_v$ say, since it is now certainly a free electron. And from Fig. 12.4, we have $\varepsilon_n(\boldsymbol{k}) - \varepsilon_F = \varepsilon_w + \varepsilon_v$, and $\varepsilon_m(\boldsymbol{k}) - \varepsilon_F = \varepsilon_w + \varepsilon_v - \hbar\omega$, so that from ε_v we can find the energy of the electron in its initial state. Thus, if we knew $\boldsymbol{k}$ for the initial state, we should know $\varepsilon_m(\boldsymbol{k})$ for that state, and by a series of such measurements we could determine the whole band structure of the metal, at least for $\varepsilon_m(\boldsymbol{k}) < \varepsilon_F$.

Now when the electron absorbs the photon, $\boldsymbol{k}$ remains unchanged, as before, because $\boldsymbol{k}_\phi$ is so small (eqn (12.13)), and when it passes out through the surface, the component of $\boldsymbol{k}$ parallel to the surface remains unchanged (just as when a light wave passes from one medium to another), so that we can write $k_\parallel = k_{v,\parallel}$. Since $k_{v,\parallel}$ is known, this gives us $k_\parallel$, and it only remains to find $k_\perp$. This is more difficult, but one possible approximation is to treat the electron in state $\varepsilon_n(\boldsymbol{k})$, inside the metal, as if it were a *free* electron of kinetic energy $\varepsilon_s + \varepsilon_v$, where ε_s is measured from the bottom of the conduction band, and so to find $k_\perp$ from $\varepsilon_s + \varepsilon_v = \hbar^2(k_\parallel^2 + k_\perp^2)/2m$. This apparently crude NFE approximation often works quite well, if $\varepsilon_s + \varepsilon_v$ is large enough ($\gtrsim 40\,\text{eV}$), because the effect of the lattice potential on such high-energy states is indeed quite small. Better methods are available when this one is too crude, but it would take us too far afield to detail them here. The essential result is that one can indeed use ARPES methods to explore $\varepsilon(\boldsymbol{k})$ in this way, by looking at the peaks in the energy spectrum of the photo-electrons emerging from the metal surface in different directions, and that the results are in good agreement with the calculated band structures, above as well as below ε_F, as shown in Fig. 12.5.

Experimentally, studies of photo-emission from metal surfaces go back to the time of Millikan and earlier, but it is only since about 1978 that they have begun to yield reasonably accurate information on $\varepsilon(\boldsymbol{k})$, for two reasons. First, the experiments need to be carried out in an extremely high vacuum, of the order of 10^{-8} Pa (10^{-10} Torr, or 10^{-13} atmosphere), if the metal surface is not to be contaminated by adsorbed gas atoms in the course of the experiment. Secondly, it is a great advantage to have available intense, tunable photon sources of energy 10–100 eV, and such sources did not exist until the advent of specially-built synchrotron radiation sources.

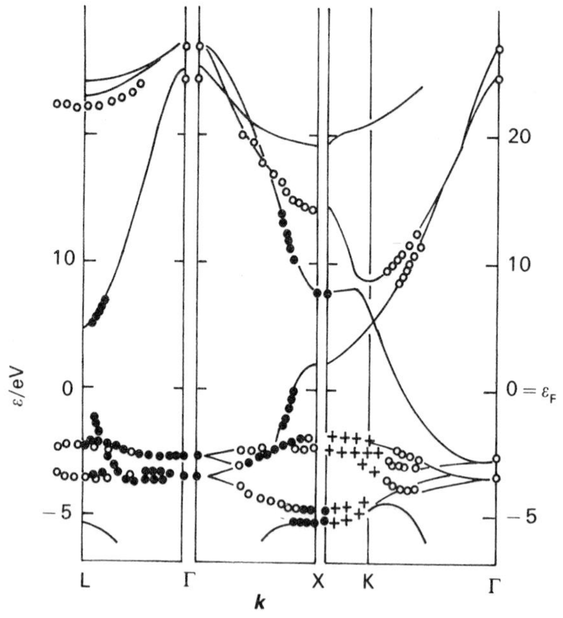

Fig. 12.5 Band structure of Cu determined by ARPES, compared with theory (full lines). [From Eastman and Himpsel (1980), *Physics of Transition Metals*, (ed. Rhodes), *Inst. of Phys. Conf. Series*, No. 55, London 1981, p. 115.]

We saw in section 7.4 how the de Haas–van Alphen effect can give us a great deal of information about the shape of the Fermi surface and about the cyclotron mass and the relaxation time of electrons on the FS; and we have seen in section 12.1 how this information can be supplemented by other measurements at radio and microwave frequencies. But these techniques tell us only about the electrons close to the FS, and moreover they work only if the electrons have rather long mean free paths – they are quite inapplicable to concentrated alloys, for example. Photo-emission studies not only enable us to study $\varepsilon(\boldsymbol{k})$ away from the FS; equally important, they also enable us to look at alloys as well as at pure metals.

We look finally at one other technique which, although not so powerful as ARPES, is also capable of looking at alloys – positron annihilation. If a positron from a radioactive source enters a metal, it rapidly loses energy, through collisions, until it is practically at rest; because there are no other positrons present, it is not prevented from doing so by the exclusion principle, as an electron would be. After a short time, the positron, by now in thermal equilibrium with the crystal and thus having energy $\sim kT$, annihilates with one of the conduction electrons, and in the process two 0.5 MeV γ-rays are emitted, which emerge from the crystal in opposite directions, or nearly so. Not quite, because the two γ-rays have to carry off between them not only the total energy of the electron and the positron, including their rest-mass energy $2m_0c^2$, but also their momentum. Almost all this momentum will be that of the electron, because the thermalized positron has so little by comparison. Consequently, the sum of the momenta $\hbar\boldsymbol{k}_1$ and $\hbar\boldsymbol{k}_2$ of the two γ-rays, $\hbar(\boldsymbol{k}_1+\boldsymbol{k}_2)=\hbar\boldsymbol{k}_e$ say, will be a measure of the momentum of the annihilated electron.

Note that $\hbar\boldsymbol{k}_e$ will not necessarily be equal to $\hbar\boldsymbol{k}$, if the electron was in state $\boldsymbol{k}$, because $\hbar\boldsymbol{k}$ is not a true momentum, only a 'crystal momentum' (cf. p. 87). Using (4.11), we can write the electron wave-function $\psi=u_{\boldsymbol{k}}(\boldsymbol{r})\exp(i\boldsymbol{k}\cdot\boldsymbol{r})$ as a sum of plane waves, $\sum_K c_K \exp[i(\boldsymbol{k}-\boldsymbol{K})\cdot\boldsymbol{r}]$, with momentum eigenvalues $\hbar(\boldsymbol{k}-\boldsymbol{K})$, and this means that a measurement of the electron momentum may yield any of the values $\hbar(\boldsymbol{k}-\boldsymbol{K})$, with probability c_K^2. But if we confine attention to small values of $\boldsymbol{k}_e$, we can put $\boldsymbol{K}=0$, and then $\boldsymbol{k}_e=\boldsymbol{k}$.

If we write $\boldsymbol{k}_e=\boldsymbol{k}_{e,\parallel}+\boldsymbol{k}_{e,\perp}$, where $\boldsymbol{k}_{e,\parallel}$ is parallel to $\boldsymbol{k}_1$ and $\boldsymbol{k}_{e,\perp}$ is normal to $\boldsymbol{k}_1$, it follows that since $k_e \ll k_1, \boldsymbol{k}_{e,\parallel}$ will make $\boldsymbol{k}_1$ and $\boldsymbol{k}_2$ slightly different in length, and $\boldsymbol{k}_{e,\perp}$ will produce a slight angle between them (Fig. 12.6). Thus if we bombard a metal sample with

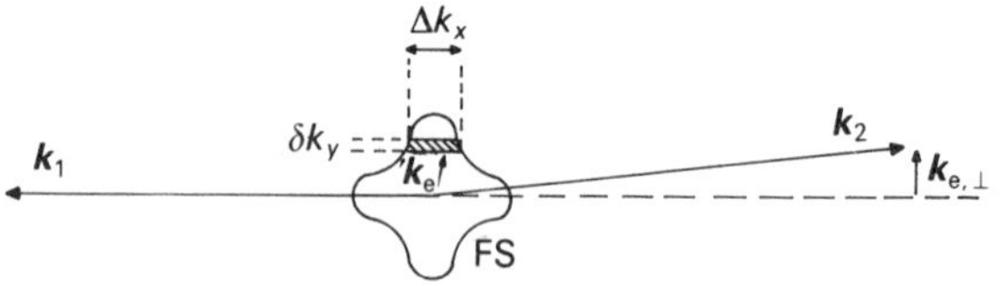

Fig. 12.6 Illustrating positron annihilation. An electron in state $\boldsymbol{k}_e$, annihilating with a thermalized positron having $\boldsymbol{k} \approx 0$, will produce two γ-rays of $\boldsymbol{k}$-vector $\boldsymbol{k}_1$ and $\boldsymbol{k}_2$, where $\boldsymbol{k}_1 + \boldsymbol{k}_2 = \boldsymbol{k}_e$. The rate of coincidences between counters set to observe γ-rays $\boldsymbol{k}_1$ and $\boldsymbol{k}_2$ will depend on the volume of the shaded region, of width Δk_x.

positrons, and set up two counters to detect coincident γ-rays emerging in the (almost opposite) directions $\boldsymbol{k}_1$ and $\boldsymbol{k}_2$, the rate at which such coincidences occur will be a measure of the number of electrons in the sample whose momentum, in the direction normal to $\boldsymbol{k}_1$, is $\hbar \boldsymbol{k}_{e,\perp}$.

In principle, we could also measure the component of the electron momentum *parallel* to $\boldsymbol{k}_1$ by measuring the difference in energy $\varepsilon_1 - \varepsilon_2 = \Delta\varepsilon$ of the two γ-rays, since $\Delta\varepsilon = \hbar c k_{e,\parallel}$, but this has seldom been attempted. We cannot hope to measure the *energy* of the electron, unfortunately, from $\varepsilon_1 + \varepsilon_2 - 2m_0c^2$, because it is far too small compared with $2m_0c^2$ to be detectable experimentally.

To see what we learn from positron annihilation experiments, consider for example the FS shown in Fig. 12.6. If we consider a small region of the BZ between k_y and $k_y + \delta k_y, k_z$ and $k_z + \delta k_z$, the filled states will occupy a volume $\Delta k_x \delta k_y \delta k_z$, where Δk_x is the distance between the two sides of the FS at k_y, k_z. Thus if the detectors are set to observe coincidences from electrons with transverse momentum components in the range k_y to $k_y + \delta k_y, k_z$ to $k_z + \delta k_z$, the measured coincidence rate will be determined by Δk_x, the length of the chord spanning the FS at k_y, k_z. Clearly, as k_y or k_z increases, Δk_x will shrink, and fall to zero when the chord passes out through the FS. We can thus deduce the size and shape of the FS, and positron annihilation experiments on Cu, for example, do indeed confirm the shape shown in Fig. 5.7. The importance of positron annihilation experiments is that, like ARPES experiments, they can be carried out on alloys as well as on pure metals.

13

Carriers in semiconductors

13.1 THE NUMBER OF CARRIERS

As we saw in section 5.2, semiconductors and insulators differ from metals in having an *energy gap* between the filled valence bands below the Fermi energy ε_F and the empty conduction bands above ε_F (Fig. 5.9). Because of this gap, the number of carriers (electrons and holes) varies rapidly with temperature in a semiconductor, whereas it is essentially temperature-independent in a metal, and our first task is to look at this temperature variation.

In practice, semiconductors almost invariably have additional 'donor' and 'acceptor' states in the gap, produced by deliberately doping the material with selected impurities, and these can have a profound effect on the properties, as we shall see later. First, though, we consider the properties of a pure semiconducting material, with no doping – the 'intrinsic' properties, as they are called.

If the energy is ε_v at the top of the valence band, and ε_c at the bottom of the conduction band, the energy gap will be $\varepsilon_g = \varepsilon_c - \varepsilon_v$. At temperatures above $T = 0$, some electrons will be thermally excited across the gap into states near the bottom of the conduction band, leaving behind an equal number of holes, i.e. unfilled states, near the top of the valence band. The probability of a state in the conduction band being filled is given as usual by the Fermi–Dirac function $f_0 = 1/(1 + e^{\eta})$, where $\eta = (\varepsilon - \varepsilon_F)/kT$, and the probability of a state in the valence band being empty is given by $1 - f_0, = 1/(e^{-\eta} + 1)$.

Now the gap width ε_g is typically about 1 eV, much larger than kT, so that if ε_F lies somewhere near the middle of the gap, only the extreme tail of the function f_0 will extend into the conduction band,

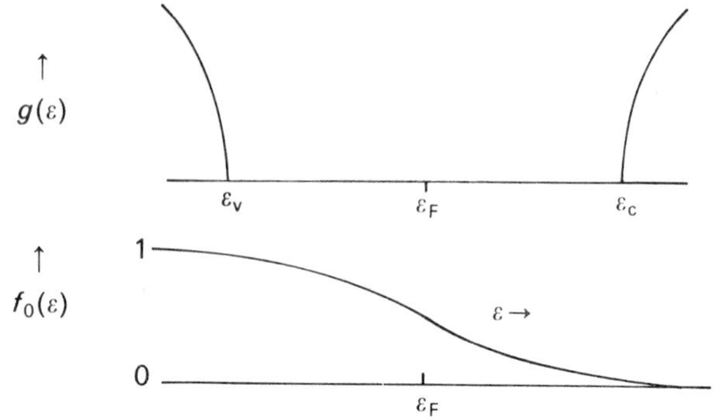

Fig. 13.1 In a semiconductor, only the tail of f_0 extends into the conduction band, and only the tail of $1-f_0$ into the valence band.

and only the extreme tail of the function $1 - f_0$ will extend into the valence band, as shown in Fig. 13.1. Within the conduction band, $e^{\eta} \gg 1$, and $f_0 = e^{-\eta}$ to good approximation, and within the valence band, $e^{-\eta} \gg 1$ and $1 - f_0 = e^{\eta}$. If the density of states is $g(\varepsilon)$, the number of electrons n_e in the conduction band and the number of holes n_h in the valence band, per unit volume, are thus given by

$$n_e = \int_{\varepsilon_c}^{\infty} g(\varepsilon)\, e^{-(\varepsilon - \varepsilon_F)/kT}\, d\varepsilon; \quad n_h = \int_{-\infty}^{\varepsilon_v} g(\varepsilon)\, e^{-(\varepsilon_F - \varepsilon)/kT}\, d\varepsilon \qquad (13.1)$$

where the exponentials become negligibly small within a few kT of the band edge, so that we can extend the integrals to infinity without appreciable error.

When f_0 (or $1 - f_0$) reduces to a Boltzmann-like exponential tail in this way, the carriers are said to be *non-degenerate*, and they behave very much like the classical Drude–Lorentz gas of Chapter 1. We are no longer concerned with properties at the Fermi surface, as we are in a metal; in a semiconductor, there *is* no Fermi surface, because ε_F lies in the energy gap.

To see the form of $g(\varepsilon)$, consider first the conduction band. If the lowest energy state, ε_c, lies at some point $\boldsymbol{k}_0$ in the Brillouin zone, then a Taylor series expansion of $\varepsilon(\boldsymbol{k}) - \varepsilon_c$ about $\boldsymbol{k}_0$ must start with quadratic terms (otherwise $\boldsymbol{k}_0$ would not be the minimum), and because we are only concerned with a region of $\boldsymbol{k}$-space very close to $\boldsymbol{k}_0$ (within kT or so in energy), we need not go beyond these terms. By taking $\boldsymbol{k}_0$ as origin, and choosing appropriate axes, we can thus write $\varepsilon(\boldsymbol{k})$ in the form

$$\varepsilon(\boldsymbol{k}) = \varepsilon_c + \alpha_x k_x^2 + \alpha_y k_y^2 + \alpha_z k_z^2$$

where $\alpha_x, \alpha_y, \alpha_z$ are constants, or (if we put $\alpha_i = \hbar^2/2m_i^*$)

$$\varepsilon(\boldsymbol{k}) = \varepsilon_c + \hbar^2(k_x^2/2m_x^* + k_y^2/2m_y^* + k_z^2/2m_z^*) \tag{13.2}$$

(cf. (7.4)). The constant-energy surfaces are thus ellipsoids, with axes $k_{\varepsilon,i} = [2m_i^*(\varepsilon - \varepsilon_c)]^{1/2}/\hbar$ $(i = x, y, z)$. The volume of such an ellipsoid will be $V_k = 4\pi k_{\varepsilon,x} k_{\varepsilon,y} k_{\varepsilon,z}/3$, and the number of states it contains (per unit volume of real space) will be $N(\varepsilon) = V_k/4\pi^3$, from (1.23). The conduction-band density of states $g_e(\varepsilon) = \mathrm{d}N(\varepsilon)/\mathrm{d}\varepsilon$ is thus given by

$$g_e(\varepsilon) = \frac{1}{2\pi^2}\left(\frac{2m_e^*}{\hbar^2}\right)^{3/2}(\varepsilon - \varepsilon_c)^{1/2} \tag{13.3}$$

where $m_e^* = (m_x^* m_y^* m_z^*)^{1/3}$. The result (13.3) is identical in form with the free-electron expression (1.26), except for the replacement of m by the effective mass m_e^*.

Exactly similar arguments applied to the valence band show that $\varepsilon(\boldsymbol{k})$ should have the form

$$\varepsilon(\boldsymbol{k}) = \varepsilon_v - \hbar^2(k_x^2/2m_{h,x}^* + k_y^2/2m_{h,y}^* + k_z^2/2m_{h,z}^*) \tag{13.4}$$

(since ε_v is now an energy *maximum*), so that

$$g_h(\varepsilon) = \frac{1}{2\pi^2}\left(\frac{2m_h^*}{\hbar^2}\right)^{3/2}(\varepsilon_v - \varepsilon)^{1/2} \tag{13.5}$$

where $m_h^* = (m_{h,x}^* m_{h,y}^* m_{h,z}^*)^{1/3}$. If we now insert these expressions in (13.1) and carry out the integrations, we find that

$$n_e = N_e\,\mathrm{e}^{-(\varepsilon_c - \varepsilon_F)/kT}; \quad n_h = N_h\,\mathrm{e}^{-(\varepsilon_F - \varepsilon_v)/kT} \tag{13.6}$$

where

$$N_e = \frac{1}{4}\left(\frac{2m_e^* kT}{\pi\hbar^2}\right)^{3/2}; \quad N_h = \frac{1}{4}\left(\frac{2m_h^* kT}{\pi\hbar^2}\right)^{3/2} \tag{13.7}$$

In deriving these results, we have nowhere assumed that $n_e = n_h$, and indeed the results are still valid in doped semiconductors, when $n_e \neq n_h$. Clearly, from (13.6), the ratio n_e/n_h will depend on the position of the Fermi energy ε_F relative to the band edges ε_c and ε_v. But if we form the product $n_e n_h$ from (13.6), we find

$$n_e n_h = N_e N_h e^{-\varepsilon_g/kT} \tag{13.8}$$

independent of ε_F, and this important relation again remains valid in doped material, when $n_e \neq n_h$. In pure material, we must have $n_e = n_h$, $= n_i$ say, where n_i is the 'intrinsic' carrier density. Then (13.8) shows that

$$n_i = (N_e N_h)^{1/2} e^{-\varepsilon_g/2kT} \tag{13.9}$$

and we can use this to calculate the electron and hole densities as a function of temperature. By combining (13.9) with (13.6), we can also find the value of ε_F in pure material:

$$\varepsilon_F = \varepsilon_v + \tfrac{1}{2}\varepsilon_g + \tfrac{1}{2}kT \ln (N_h/N_e) \tag{13.10}$$

Thus, unless N_h and N_e are very different, ε_F will lie within kT or so of the middle of the band gap.

To apply to a real semiconductor such as Si, equations (13.7) need generalizing slightly. We saw in section 5.2 that in Si there are, by symmetry, six equivalent minima in the conduction band, each located at a different point $\boldsymbol{k}_0$, so that N_e is in fact the sum of six equal contributions of the form (13.7). There are also three valence bands, all having their maxima at $\boldsymbol{k} = 0$, so that by symmetry we would expect their energy surfaces to be spherical. In fact two of them perturb each other apart to form 'warped' spheres, but we can neglect this complication and write

$$\varepsilon(\boldsymbol{k}) = \varepsilon_v - \hbar^2 k^2/2m^*_{h,1}; \quad \varepsilon(\boldsymbol{k}) = \varepsilon_v - \hbar^2 k^2/2m^*_{h,2} \tag{13.11}$$

for the heavy holes (of mass $m^*_{h,1}$) and light holes (of mass $m^*_{h,2}$) respectively. The third band lies somewhat lower than the other two, with $\varepsilon(\boldsymbol{k})$ given by

$$\varepsilon(\boldsymbol{k}) = \varepsilon_v - \varepsilon_3 - \hbar^2 k^2/2m^*_{h,3} \tag{13.12}$$

where $m^*_{h,3}$ is intermediate between $m^*_{h,1}$ and $m^*_{h,2}$. Because of the energy gap ε_3 between this band and the others, it will contain few holes and can usually be neglected. Thus N_h is the sum of two terms of the form (13.7), one from the heavy holes and one from the light holes (problem 13.1).

13.2 DONORS AND ACCEPTORS

Most metals contain something like one or two conduction electrons per atom. By contrast, problem 13.1 shows that in ideally pure Si at 300 K, only about one electron per 10^{13} atoms is thermally excited into the conduction band, leaving behind a hole in the valence band. It is therefore not surprising that minute amounts of impurity – at the level of parts per million or less – can have a very large effect on the properties of semiconductors, if the impurity atoms tend to release electrons into the conduction band, or capture them from the valence band. This is precisely what donor and acceptor atoms do.

Si and Ge have a chemical valency of 4; that is, the free atom has four electrons outside the ion core, and it is these that fill the valence bands in the semiconducting crystal. If an impurity atom with a valency of 5, such as P or As, is substituted for one of the Si or Ge atoms in the crystal, it will have one extra electron, which it can rather easily lose to the conduction band, thus becoming 'ionized'; the additional positive charge on the impurity ion is then no longer screened by the extra electron. Likewise, a trivalent impurity such as Ga or In has one less electron than its neighbours, and can rather easily capture one from the valence band, creating a hole there which can move away through the crystal. Again this leaves the impurity 'ionized', this time negatively. Thus P and As act as donors, and Ga and In as acceptors.

How much energy is needed to 'ionize' a donor or acceptor? When a donor is ionized, it becomes positively charged relative to the surrounding ions, and will exert a Coulomb attraction on any electrons in the conduction band – just as a proton in vacuum would attract a passing electron. And just as a proton can capture a passing electron to form a hydrogen atom, so the ionized donor can capture an electron from the conduction band to become non-ionized. The two processes are very similar, and the energy needed to ionize a donor atom can be calculated by analogy with the ionization energy of a hydrogen atom. That energy is given by

$$\varepsilon_i = e^2/8\pi\varepsilon_0 r_0 = me^4/32\pi^2\varepsilon_0^2\hbar^2 \qquad (13.13)$$

where $r_0 = 4\pi\varepsilon_0\hbar^2/me^2$ is the Bohr radius. In the present problem, the effective mass of the electron is not the free mass m but some combination of the masses m_x^*, m_y^* and m_z^* of (13.2), which are typically around $0.2m$, and this will reduce the ionization energy, typically by a factor of five or so.

More important, the electron is now moving through a medium with a rather high dielectric constant ε_r, about 12 in Si, and to allow for this we must replace ε_0 by $\varepsilon_0\varepsilon_r$ in (13.13), which reduces ε_i by a further factor of 150 or so. (*Do not* confuse the dielectric constant ε_0 or ε_r here with an energy!) The net result is that the ionization energy ε_d of a donor in a semiconductor is far less than the 13.6 eV given by (13.13): for P in Si, for example, it is 0.044 eV, and for P in Ge only 0.012 eV. Correspondingly, the effective Bohr radius for an electron bound to the donor atom is now given by $r_{\text{eff}} = 4\pi\varepsilon_r\varepsilon_0\hbar^2/m^*e^2$, 25–50 Å instead of 0.5 Å (i.e. 2.5–5 nm instead of 0.05 nm), so that the wave-function extends over many unit cells; this is why we are able to treat the crystal, as we have done, as simply a continuous medium with a high dielectric constant.

In exactly the same way, we can think of an ionized acceptor, negatively charged relative to the surrounding ions, as attracting and capturing a passing hole from the valence band to become non-ionized. The energy needed to ionize it again, by detaching the hole and setting it free in the valence band (in other words, by capturing an electron from the valence band), is again given by replacing m by an effective hole mass m_h^* in (13.13), and replacing ε_0 by $\varepsilon_0\varepsilon_r$. The effect is again to make the acceptor ionization energy ε_a very small, about 0.070 eV for Ga in Si, and 0.011 eV for Ga in Ge. (Some impurities turn out to produce donor or acceptor levels with much larger values of ε_d or ε_a, but we shall not consider these 'deep levels' here.)

We can thus draw an energy level diagram, Fig. 13.2, showing the donor levels a distance ε_d below the bottom of the conduction band and the acceptor levels a distance ε_a above the top of the valence band. It takes an energy ε_d to raise an electron from a donor level into the conduction band, leaving the donor ionized; it takes an energy ε_a to lower a hole from an acceptor level into the valence band (i.e. to raise an electron from the valence band to the acceptor level), leaving the acceptor ionized. Because ε_d and ε_a are so small – comparable with or smaller than kT at room temperature – the donors and acceptors will be largely ionized at normal temperatures. The probability that a donor state is occupied by an electron

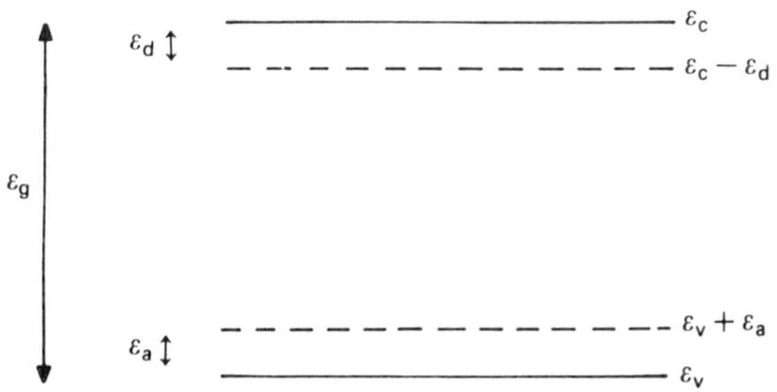

Fig. 13.2 Energy levels of donors and acceptors, relative to the conduction band edge ε_c and valence band edge ε_v.

is given, to good approximation, by the factor $f_0(\varepsilon)$, with $\varepsilon = \varepsilon_c - \varepsilon_d$, and the probability that an acceptor state is occupied by a hole is given by $1 - f_0(\varepsilon)$, with $\varepsilon = \varepsilon_v + \varepsilon_a$. (In fact the statistics of a donor or acceptor state are not quite the same as the Fermi–Dirac statistics of an electron gas. The probability of a donor state of energy ε being occupied is $1/[1 + \frac{1}{2}\exp(\varepsilon - \varepsilon_F)/kT]$, and the probability of an acceptor state of energy ε being occupied by a hole is $1/[1 + \frac{1}{2}\exp(\varepsilon_F - \varepsilon)/kT]$. But the factors of $\frac{1}{2}$ here can usually be ignored (problem 13.2) and the probabilities then become f_0 and $1 - f_0$.)

Any real semiconductor is likely to contain both donors and acceptors – say N_d donors and N_a acceptors per unit volume. If $N_d > N_a$, the material is said to be *n-type*, because there will be more electrons than holes, so that the *majority carriers* are negative electrons; if $N_d < N_a$, the material is *p-type*, because the majority carriers are positive holes. (If $N_d \sim N_a$, the material is said to be *compensated*.) Now, if an electron occupies a donor state at energy $\varepsilon_c - \varepsilon_d$, and a hole occupies an acceptor state at energy $\varepsilon_v + \varepsilon_a$, the electron can lower its energy by $(\varepsilon_c - \varepsilon_d) - (\varepsilon_v + \varepsilon_a) = \varepsilon_g - (\varepsilon_d + \varepsilon_a)$ by dropping into the acceptor state, leaving both states ionized. At $T = 0$, all the N_a acceptor states and N_a of the donor states will therefore be ionized (if $N_d > N_a$), leaving $(N_d - N_a)$ donor states still occupied by electrons. The conduction band will be completely empty, and the valence band completely full, so that $n_e = n_h = 0$. The Fermi energy ε_F must then coincide with the donor level, because the donor states are only partly filled.

As T is raised, the electrons will redistribute themselves among the various levels, but the total number of electrons must remain unchanged. To see what this implies, let N_v be the total number of states in the valence band, and let $N_{a,h}$ $(=N_a(1-f_0))$ and $N_{d,e}$ $(=N_d f_0)$ be the number of holes in the acceptor states and the number of electrons in the donor states. The situation at $T=0$ and at $T>0$ is then summarized in the table:

Electrons in:	Valence band	Acceptor states	Donor states	Conduction band
$T=0$	N_v	N_a	$N_d - N_a$	0
$T>0$	$N_v - n_h$	$N_a - N_{a,h}$	$N_{d,e}$	n_e

If the total number of electrons is to be independent of temperature, it follows that we must have

$$n_e + N_{d,e} = n_h + N_{a,h} + (N_d - N_a) \tag{13.14}$$

and it is easy to see that this must also hold for p-type material (with $N_d < N_a$). For n-type material, (13.14) reduces to $N_{d,e} = N_d - N_a$ at $T=0$, as expected, because all other terms then vanish. For $T>0$, we can distinguish two temperature regions: the *extrinsic* region, $T<T_i$ say, where n_h and $N_{a,h}$ remain negligible, and the *intrinsic* region, $T>T_i$, where n_h becomes comparable with n_e.

Since $n_e n_h = n_i^2$ from (13.8), n_h will be negligible as long as $n_e \gg n_i$. In exactly the same way we can show that $N_{d,e}N_{a,h} = N_i^2$, where $N_i = (N_d N_a)^{1/2} e^{-(\varepsilon_g - \varepsilon_d - \varepsilon_a)/2kT}$, (problem 13.2), so that $N_{a,h}$ will be negligible as long as $N_{d,e} \gg N_i$. In the extrinsic region, then, (13.14) reduces to $n_e + N_{d,e} = N_d - N_a$. Here n_e is given by (13.6), and $N_{d,e}$ by the corresponding expression $N_{d,e} = N_d e^{-(\varepsilon_c - \varepsilon_d - \varepsilon_F)/kT}$, which is valid as long as the function f_0 can be approximated by its exponential tail (and is thus valid down to $T=0$, if $N_d - N_a \gtrsim 0.2N_d$). We thus find (as we might expect) that the $N_d - N_a$ electrons available divide themselves between the donor states and the conduction band in the ratio $N_d : N_e e^{-\varepsilon_d/kT}$, so that

$$n_e = (N_d - N_a)/(1 + N_d e^{\varepsilon_d/kT}/N_e) \tag{13.15}$$

Now at room temperature, $N_e \sim 10^{25}\,\mathrm{m}^{-3}$ (problem 13.1), while the donor concentration N_d will seldom exceed $10^{24}\,\mathrm{m}^{-3}$, so that (13.15) reduces to $n_e = N_d - N_a$ (in other words, the donors are fully ionized)

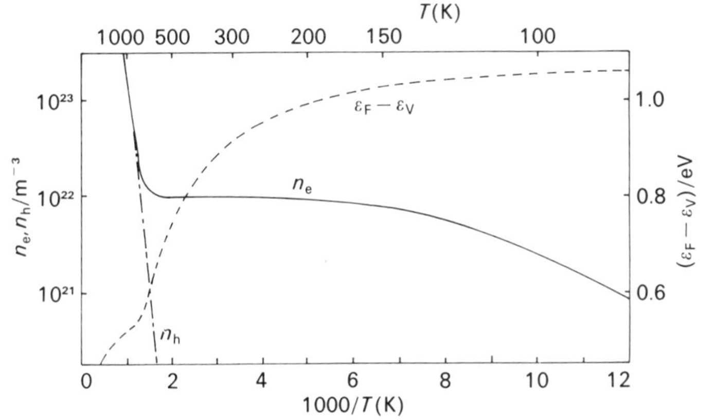

Fig. 13.3 Variation of n_e, n_h and $\varepsilon_F - \varepsilon_v$ with temperature in a doped n-type Si sample.

except at low temperatures, where $kT \ll \varepsilon_d$ and n_e falls exponentially to zero.

Except at low temperatures, then, the conduction band density n_e stays constant as long as n_h and $N_{a,h}$ can be neglected in (13.14). But as the temperature rises, the product $n_e n_h$ grows exponentially, from (13.8), and at $T \sim T_i$, n_h becomes comparable with n_e. For $T > T_i$, both n_e and n_h are large compared with $N_d - N_a$, and also large compared with $N_{d,e}$ and $N_{a,h}$, so that (13.14) reduces effectively to $n_e = n_h = n_i$. The properties of the material are then no longer affected by the doping concentrations N_d and N_a, but become the same as those of pure material – the *intrinsic* properties. Finally, to find how n_e varies for $T \approx T_i$, we simply have to write $n_h = n_i^2/n_e$ and $N_{a,h} = N_i^2/N_{d,e}$ in (13.14), and solve the resultant quadratic equation.

The variation of n_e, n_h and ε_F with temperature is shown in Fig. 13.3 for a hypothetical Si sample with $N_d - N_a = 10^{22}\,m^{-3}$, assuming $\varepsilon_d = 0.044\,eV$. For this sample $T_i \approx 750\,K$, and above this temperature $n_e \approx n_h$, and ε_F is given by (13.10). Between about 700 K and 150 K, $n_e \approx N_d - N_a$, and below 150 K n_e falls exponentially to zero as predicted by (13.15), while ε_F rises to $\varepsilon_c - \varepsilon_d$. For a p-type sample the curves for n_e and n_h would be interchanged, and the curve for ε_F would tend down towards $\varepsilon_v + \varepsilon_a$ at low temperatures, rather than up to $\varepsilon_c - \varepsilon_d$.

13.3 CARRIER MOBILITIES AND POSITIVE HOLES

Although the same basic physics determines the transport properties of both metals and semiconductors, the detailed treatment is somewhat different in the two cases, because the Fermi surface in a metal is much larger and usually more complex in shape than the relevant energy surfaces in semiconductors. Consequently, the reciprocal mass tensor (7.4) is seldom useful in discussing metals, because the components m_{ij}^* vary from point to point on the FS, whereas the relevant energy surfaces in a semiconductor – those within kT or so of the band edge – are likely to be well approximated by ellipsoids of the form (13.2) or (13.4), so that the reciprocal mass tensor is a much more useful concept.

Again, the effect of collisions in metals is normally described in terms of the relaxation time τ or the mean free path l, whereas in a semiconductor one more often uses a related quantity, the *mobility* $\mu = e\tau/m^*$, which describes the drift velocity acquired by the carriers in unit applied electric field. In metals, too, we seldom have to worry about the variation of τ_k with the energy ε_k of the electron, because it varies only slightly over the relevant range of energies, $\varepsilon_F \pm kT$. In semiconductors, on the other hand, the relevant range of energies extends from *zero* (relative to the band edge) to a few kT, and τ_k may vary widely within this range. We therefore have to allow for this in finding the transport properties.

Lastly, in discussing metals, we had no need to introduce the concept of *positive holes*. We saw that orbits on the FS in a magnetic field could be classified into electron orbits and hole orbits, and that on this basis the surfaces themselves could be classed as either electron surfaces or hole surfaces, but this did not affect the expressions for the transport properties, which were determined by the same integrals over the FS in both cases. But in discussing semiconductors, it is convenient to think of the valence band as if it were an almost empty band containing a few positively charged carriers – holes – near the top of the band, instead of an almost full band containing a few vacant states near the top.

Since a full band can carry no current, the current carried by an almost full band must be equal and opposite to that which *would* have been carried by the missing electrons. So we can replace the full band, containing a few missing electrons, by an empty band containing a few *positive* holes. This will produce the current we want, provided that the holes respond to an applied field as the missing electrons would have done. Now the effective mass of the electrons near the top of the band is *negative*; as we can see by comparing (13.4) with (7.4), the mass components are $-m^*_{h,x}$, $-m^*_{h,y}$ and $-m^*_{h,z}$. If follows that positive carriers, having charge $+e$, will respond to an applied field $\boldsymbol{E}$ in exactly the same way as these electrons, provided that we give the positive carriers *positive* mass components $m^*_{h,x}$, $m^*_{h,y}$ and $m^*_{h,z}$. Because of their opposite charge, they will then carry a current exactly equal and opposite to that which would have been carried by the missing electrons – and that is just what we want. Moreover, the positive mass of these carriers also makes sense in energy terms. If the filled valence band is taken as having zero energy, then the energy when an electron is missing from state $\boldsymbol{k}$ will be

$$\varepsilon(\boldsymbol{k}) = -\varepsilon_v + \hbar^2(k_x^2/2m^*_{h,x} + k_y^2/2m^*_{h,y} + k_z^2/2m^*_{h,z}) \qquad (13.16)$$

and if we take this as representing the energy of the positive hole, then it will indeed have positive mass components.

We can thus think of the valence band of a semiconductor as containing n_h positive holes, each carrying charge $+e$ and having positive effective mass. We have indeed already made use of this picture, in discussing the ionization energy of acceptor states, and we can now use it to discuss the transport properties.

14

Transport properties of semiconductors

14.1 SCATTERING

Electrons (or holes) in semiconductors, just like electrons in metals, can be scattered either by static defects or by phonons. As in metals, phonon scattering will be dominant at high temperatures, but dies away at low temperatures, and scattering is then dominated by the static imperfections, in this case mainly the ionized impurities. We have already dealt with scattering by charged impurities in metals in sections 8.1 and 9.2, and exactly the same treatment applies to semiconductors, except that in (8.7) and (8.11) we should replace ε_0 by $\varepsilon_0\varepsilon_r$ to allow for the dielectric constant of the material. The effective relaxation time τ^e is still given by (9.14), and if we assume for simplicity that the energy surfaces are spherical, with $\varepsilon_k = \hbar^2k^2/2m^* = \frac{1}{2}m^*v^2$, the matrix element $V_{\boldsymbol{kk'}}$ is given by $V_{kk'} = \Delta Z e^2 r_1^2/\varepsilon_r\varepsilon_0(1 + k_s^2 r_1^2)$, as in problem 8.2. Putting $\Delta Z = \pm 1$ for singly charged donors or acceptors, and putting $k_s = 2k\sin\frac{1}{2}\theta$, $\mathrm{d}S_{k'} = 2\pi k^2 \sin\theta\,\mathrm{d}\theta$, (9.14) becomes

$$\frac{1}{\tau_k^e} = \frac{N_I}{8\pi}\left(\frac{e^2}{\varepsilon_r\varepsilon_0}\right)^2 \frac{1}{(2m^*)^{1/2}\varepsilon_k^{3/2}} \int_0^1 \frac{z^3\,\mathrm{d}z}{(z^2 + 1/\alpha^2)^2} \tag{14.1}$$

where N_I is the number of ionized impurities per unit volume, $z = \sin\frac{1}{2}\theta$ and $\alpha = 2kr_1$, with r_1 given by (8.10). The integral is equal to $\frac{1}{2}[\ln(1+\alpha^2) - \alpha^2/(1+\alpha^2)]$, and since normally $\alpha^2 \gg 1$ (problem 14.1), this reduces to $\ln\alpha - \frac{1}{2}$, so that finally we have

$$\frac{1}{\tau_k^e} = \frac{N_I}{8\pi}\left(\frac{e^2}{\varepsilon_r\varepsilon_0}\right)^2 \frac{\ln\alpha - \frac{1}{2}}{(2m^*)^{1/2}\varepsilon_k^{3/2}}. \tag{14.2}$$

This is the so-called Brooks–Herring expression for τ^e, and it applies equally well to the scattering of electrons or holes; in either case, ε_k is the energy of the carrier as measured from the band edge.

An alternative approximate treatment of ionized impurity scattering, due to Conwell and Weisskopf, neglects the effect of screening and so puts $r_1 = \alpha = \infty$. The integral would then diverge, and to avoid this the lower limit is replaced by z_{min}. The value of z_{min} is fixed by arguing that the impurity ions are separated by a distance of order $N_I^{-1/3}$, so that the minimum possible scattering angle θ_{min} (and hence z_{min}) occurs when an electron passes at a distance $b_{min} = \frac{1}{2}N_I^{-1/3}$ from the ion. Here b_{min} is related to θ_{min} by the usual Rutherford scattering formula, $b = Ze^2 \cot\frac{1}{2}\theta/8\pi\varepsilon_r\varepsilon_0\varepsilon_k$. The integral then reduces to $\ln(1/\sin\frac{1}{2}\theta_{min})$, $= \frac{1}{2}\ln(1 + \cot^2\frac{1}{2}\theta_{min})$, so that the net effect is simply to replace $\ln\alpha - \frac{1}{2}$ in (14.2) by $\frac{1}{2}\ln(1 + \alpha_1^2)$, with $\alpha_1 = 4\pi\varepsilon_r\varepsilon_0\varepsilon_k/e^2 N_I^{1/3}$. Usually $\alpha_1^2 \gg 1$ (problem 14.1), like α^2, so that $\frac{1}{2}\ln(1 + \alpha_1^2)$ reduces to $\ln\alpha_1$, and in fact this approach yields quite similar values for $1/\tau^e$ to those given by (14.2). In fact we do not expect either of these approaches to predict $1/\tau^e$ exactly, because of the various approximations involved, but they both give answers which are usually correct to within a factor of 2 or 3. And they both predict that $1/\tau^e$ should vary as $\varepsilon_k^{-3/2}$, to good approximation, because the $\ln\alpha$ or $\ln\alpha_1$ term only varies rather slowly with ε_k or T. The effective scattering rate thus gets smaller as the energy increases, as we would expect, because fast-moving carriers are deflected less by the scattering centres.

Except at the lowest temperatures, there will also be phonon scattering to consider. Here again, the essentials were set out in Chapters 8 and 9. In a semiconductor, because the occupied region of $\boldsymbol{k}$-space is so small, it needs only a very low-energy phonon to reverse the direction of motion of an electron (or hole). Thus if we again write $\varepsilon_k = \hbar^2k^2/2m^*$, with ε_k measured from the band edge, and put $\varepsilon_k \approx kT$, we have $|\boldsymbol{k}| \approx (2m^*kT)^{1/2}/\hbar$. To reverse the direction of $\boldsymbol{k}$ we need a phonon of wave-vector $\boldsymbol{q} = 2\boldsymbol{k}$, and energy $\varepsilon_q = \hbar\omega_q \approx 2\hbar v_s k$, where v_s is the velocity of sound. We thus have $\varepsilon_q/\varepsilon_k = (8m^*v_s^2/kT)^{1/2}$, which for $v_s \sim 5000\,\mathrm{m\,s^{-1}}$ and $m^* \sim 0.1m$ turns out to be ~ 0.1 or so at $T = 100\,\mathrm{K}$, and less at higher temperatures. Phonons of such low energy will be fully excited down to very low temperatures, so that we can replace n_q in (9.15) by $kT/\hbar\omega_q$, and that equation becomes

$$1/\tau^{e} = (V_0^2/4\pi^2\hbar NM)\int dS_k(kTq^2/\hbar\omega_q^2 v_k)(1-\cos\theta)$$

$$= (V_0^2/4\pi^2\hbar^2 NM)(kT/v_s^2 v_k)\,4\pi k^2 \qquad (14.3)$$

since the integral is over a sphere of radius k, and the $\cos\theta$ term integrates to zero. Writing $\varepsilon_k = \frac{1}{2}m^* v_k^2 = \hbar^2k^2/2m^*$, this becomes

$$1/\tau^{e} = (V_0^2 kT/2\pi\hbar^4 NMv_s^2)(2m^*)^{3/2}\varepsilon_k^{1/2} \qquad (14.4)$$

(where, as before, NM is simply the density of the crystal). The quantity V_0 is a measure of the strength of the electron–phonon coupling, and (on a somewhat simplified model) is equal to the *deformation potential.* This measures the rate at which the energy ε_c or ε_v at the band edge varies with the volume V of the crystal: $V_0 = V\,d\varepsilon_c/dV$ or $V\,d\varepsilon_v/dV$. In a metal, we can write $V_0 \approx 2\varepsilon_F/3$, as we did below (8.18), but clearly that relation no longer holds in a semiconductor; we can either find $d\varepsilon_c/dV$ from a band-structure calculation, or just regard V_0 as an empirical parameter to describe the strength of the coupling. In semiconductors, as in metals, V_0 is typically a few eV.

Whereas the scattering rate due to ionized impurities decreased as ε_k increased, equation (14.4) shows that the phonon scattering rate increases, and this is because at higher energies there are more states available for the electron or hole to scatter into, as measured by the area $4\pi k^2$ of the suface integral $\int dS_k$ in (14.3). Since $v_k \propto \varepsilon_k^{1/2}$, (14.4) shows that the mean free path l is independent of ε_k. As in metals, the scattering rate varies as T, simply because the number of phonons varies as T.

There are various other sources of scattering in semiconductors. Not all the impurities will necessarily be ionized, and neutral impurities will also scatter, though more weakly than ionized ones. And although only phonons of very low energy are needed to produce the *intra-valley* scattering that we have so far considered, *inter-valley* scattering can also occur, if there is more than one point in the conduction band where ε takes on its minimum value ε_c. In Si, for example, there are six such points, as we have seen (and in Ge, there are four), and electrons can be scattered from one of these minima to another. But this will require phonons of much higher

$\boldsymbol{q}$, and hence higher energy, than intra-valley scattering, which is usually the dominant process at room temperature.

In semiconductors such as InSb or GaAs, intervalley scattering will not occur, because the conduction band minimum occurs at $\boldsymbol{k} = 0$, so that there is only one valley. On the other hand these crystals have no centre of symmetry and thus, unlike Si or Ge, are piezoelectric: the strain produced by a phonon will generate an electric field, and the electrons can thus be scattered by phonons through this piezoelectric coupling as well as through deformation potential coupling.

14.2 SIMPLE TRANSPORT PROPERTIES

Suppose we have n-type material, so that the carriers are electrons rather than holes. (For p-type material, precisely similar arguments will lead to precisely similar results.) Then for the ellipsoidal energy surfaces (13.2), we have $\hbar k_x = m_x^* v_x$, etc. and on the simple approach used in Chapter 1 we should expect an electric field E_x to produce a drift velocity $v_{\mathrm{d},x} = eE_x\tau/m_x^*$, so that $\sigma_{xx} = n_\mathrm{e}e^2\tau/m_x^*$. This result is qualitatively correct, but we have seen that τ varies with ε (as $\varepsilon^{3/2}$ for ionized impurity scattering, and as $\varepsilon^{-1/2}$ for phonon scattering), so that we need some appropriate average of $\tau(\varepsilon)$. To calculate σ_{xx} correctly, we start from (9.8), which is just as valid for semiconductors as for metals, and as usual we write $L_x = \tau v_x$. Then (9.8) becomes

$$\begin{aligned}\sigma_{xx} &= -(e^2/4\pi^3)\int v_x^2\tau(\partial f_0/\partial\varepsilon)\mathrm{d}^3\boldsymbol{k} \\ &= (e^2/4\pi^3 kT)\int v_x^2\tau f_0\mathrm{d}^3\boldsymbol{k} \qquad (14.5)\end{aligned}$$

if $f_0(\varepsilon)$ varies as $\mathrm{e}^{-\varepsilon/kT}$. To evaluate (14.5), we write ε (measured from the band edge) in the form $\varepsilon = \kappa_x^2 + \kappa_y^2 + \kappa_z^2$, where $\kappa_\mathrm{i} = \hbar k_\mathrm{i}/\sqrt{(2m_\mathrm{i}^*)}$, so that a constant-energy surface is a sphere in κ-space, and so that $v_x^2 = 2\kappa_x^2/m_x^*$. We also write $\tau(\varepsilon) = \tau_0 u^\nu$, where τ_0 is the value of $\tau(\varepsilon)$ at $\varepsilon = kT$ and $u = \varepsilon/kT$, so that $\nu = \frac{3}{2}$ for ionized impurity scattering, $-\frac{1}{2}$ for phonon scattering. Then after a little algebra (problem 14.2), we find that

$$\sigma_{xx} = n_\mathrm{e}e^2\langle\tau\rangle/m_x^* \qquad (14.6)$$

where n_e is the number of electrons in the ellipsoid and

$$\langle\tau\rangle = \tau_0 \int_0^\infty u^{3/2+\nu} e^{-u}\,du \Big/ \int_0^\infty u^{3/2} e^{-u}\,du \qquad (14.7)$$

so that $\langle\tau\rangle = 4.51\tau_0$ for $\nu = \frac{3}{2}$ and $0.752\tau_0$ for $\nu = -\frac{1}{2}$.

The mean conductivity $\sigma_e = (\sigma_{xx} + \sigma_{yy} + \sigma_{zz})/3$ of the ellipsoid is thus given by

$$\sigma_e = n_e e^2 \langle\tau\rangle / \langle m^*\rangle, \qquad (14.8)$$

where $1/\langle m^*\rangle = (1/m_x^* + 1/m_y^* + 1/m_z^*)/3$, and the same expression gives the total conductivity of a set of equivalent ellipsoids, if n_e is now taken to be the total number of electrons in all of them, i.e. the total number of electrons in the conduction band. In p-type material, of course, (14.8) is replaced by $\sigma_h = n_h e^2 \langle\tau_h\rangle / \langle m_h^*\rangle$, where n_h, $\langle\tau_h\rangle$ and $\langle m_h^*\rangle$ are the appropriate parameters for the valence band. In the intrinsic region, where both electrons and holes are present in comparable numbers, the total conductivity is the sum of the separate contributions σ_e and σ_h from the electrons and the holes.

If two different scattering mechanisms are present, with relaxation times $\tau_{01}u^{\nu_1}$ and $\tau_{02}u^{\nu_2}$, the total scattering rate at energy $\varepsilon = ukT$ will be $1/\tau = 1/\tau_1 + 1/\tau_2 = u^{-\nu_1}/\tau_{01} + u^{-\nu_2}/\tau_{02}$, and in principle we should find $\langle\tau\rangle$ by evaluating (14.7) with $\tau_0 u^\nu$ replaced by this expression for τ. Usually, though, it is enough to use the rough approximation $1/\langle\tau\rangle = 1/\langle\tau_1\rangle + 1/\langle\tau_2\rangle$, and in this approximation Matthiessen's rule holds for semiconductors as well as for metals.

In semiconductors, unlike metals, the temperature-dependence of σ_e is dominated by n_e. As Fig. 13.3 shows, changing the temperature T by a factor of 2 can change n_e by a factor of 10 or more. The effective relaxation time $\langle\tau\rangle$ varies much more slowly – as $T^{3/2}$ for ionized impurity scattering (from (14.2)) and as $T^{-3/2}$ for phonon scattering (from (14.4)). (Remember that $\langle\tau\rangle \propto \tau_0$, where τ_0 is the value of τ when $\varepsilon = kT$.) Consequently a plot of σ against $1/T$ has much the same shape as a plot of n_e against $1/T$, as Fig. 14.1 shows. Over the temperature range where n_e is practically constant, σ may rise or fall, depending on whether impurity scattering or phonon scattering is more important.

The actual magnitude of σ is several orders of magnitude less in semiconductors than in metals, because the number of carriers n_e or n_h is so much smaller. This more than outweighs the substantially

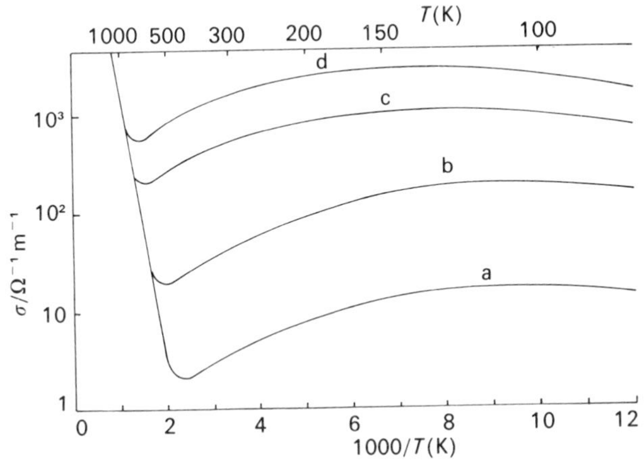

Fig. 14.1 Conductivity $\sigma(T)$ of n-type Si, doped with As. Curves (a) to (d) are for $N_d \approx 3 \times 10^{20}$, 2.5×10^{21}, 2×10^{22} and $1.3 \times 10^{23}\,\mathrm{m}^{-3}$ respectively.

higher mobility of the carriers, which may be two or three orders of magnitude greater in a semiconductor than in a metal. In Cu, for example, τ at room temperature is about $2 \times 10^{-14}\,\mathrm{s}$, so that $\mu = e\tau/m^*$ is about $3.5 \times 10^{-3}\,\mathrm{m^2\,V^{-1}\,s^{-1}}$ (so that a field of $1\,\mathrm{V\,m^{-1}}$ produces an electron drift velocity of about $3.5\,\mathrm{mm\,s^{-1}}$). In Si, on the other hand, $\mu_e \approx 0.15\,\mathrm{m^2\,V^{-1}\,s^{-1}}$ and $\mu_h \approx 0.05\,\mathrm{m^2\,V^{-1}\,s^{-1}}$ at room temperature, and in InSb μ_e approaches $10\,\mathrm{m^2\,V^{-1}\,s^{-1}}$. At least two factors combine to make these mobilities so much higher than those in metals: the effective masses m^* are usually much smaller, and the density of final states to scatter into – the area of integration in (14.3) – is also much smaller, so that τ is much larger.

To determine μ experimentally, we can divide σ by $n_e e$ for n-type or $n_h e$ for p-type material, with $n_e e$ or $n_h e$ determined by Hall effect measurements (see below). The results are then called Hall mobilities. Alternatively, μ can be measured directly for the *minority* carriers – that is, for the holes in n-type material or for the electrons in p-type material – by applying an electric field along the length of a bar of the material and injecting minority carriers into the bar near one end, which can be done by using the appropriate kind of contact, and seeing how long the carriers take to drift to a second contact further along the bar. The drift mobilites determined in this way usually agree reasonably well with the Hall mobilities.

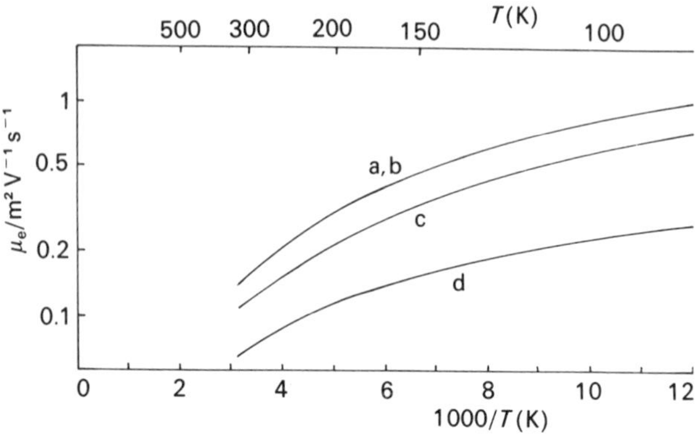

Fig. 14.2 Hall mobility $\mu(T)$ of the Si samples of Fig. 14.1 [Figs 14.1 and 14.2 after Morin and Maita (1954), *Phys. Rev.* **96**, 28.]

Figure 14.2 shows the way μ varies with temperature in various samples of Si.

Just as the conductivity is still given by (9.8) for semiconductors as well as for metals, so the thermoelectric power is still given by $S = C/\sigma$, with C given by (10.14). But for an n-type semiconductor we can usually replace $\varepsilon - \varepsilon_F$ in (10.14) by the constant quantity $\varepsilon_c - \varepsilon_F$, because the occupied states only extend from ε_c to $\varepsilon_c + kT$ or so in energy, and this range is generally small compared with $\varepsilon_c - \varepsilon_F$. If we do this, and if we neglect the difference between $\boldsymbol{L}$ and $\boldsymbol{L}^T$ – which is usually well justified in semiconductors, because inelastic small-angle scattering is usually unimportant – (10.14) or (10.15) reduces at once to $CT = -(\varepsilon_c - \varepsilon_F)\sigma/e$ (cf. (9.9)), so that $S = -(\varepsilon_c - \varepsilon_F)/eT$. Comparing this with (10.28), we see that the thermoelectric power in a semiconductor will be very much larger than in a metal. In a metal, S is smaller than k/e by a factor of order kT/ε_F; in a semiconductor, it is larger than k/e by a factor of order ε_g/kT.

The reason for the difference is clear enough. In a metal, as we saw in section 10.2, the thermoelectric effects are caused by a slight imbalance between 'hot' electrons flowing one way and 'cool' ones flowing the other way. In an n-type semiconductor there is no longer merely a slight imbalance; *all* the carriers are 'hot', in the sense that they have $\varepsilon > \varepsilon_F$, and in a p-type semiconductor *all* the carriers are 'cool', with $\varepsilon < \varepsilon_F$.

In n-type material, S will be negative, and in p-type material it will be positive. In intrinsic material, we can write $C = C_e + C_h = S_e\sigma_e + S_h\sigma_h$ where $S_e(S_h)$ and $\sigma_e(\sigma_h)$ are the thermoelectric power and conductivity due to the electrons (holes) only. But we also have $C = S\sigma = S(\sigma_e + \sigma_h)$ where S is the resultant total thermoelectric power, and it follows that $S = (S_e\sigma_e + S_h\sigma_h)/(\sigma_e + \sigma_h)$. Since S_e and S_h are of opposite sign, the resultant S may be comparatively small.

The thermal conductivity κ of a semiconductor, like σ, is much less than that of a metal. Most of the heat is usually carried by the phonons, as in an insulator, rather than by the electrons. Moreover, even after subtracting out the phonon contribution, the quantity κ_c actually measured may be a good deal smaller than κ if S is large, as (10.30) shows. As a matter of interest, one can write κ_c in a form closely similar to (10.6) for κ:

$$\kappa_c = -(1/12\pi^3 T)\int \boldsymbol{v}_{\boldsymbol{k}}\cdot\boldsymbol{L}_{\boldsymbol{k}}^T(\varepsilon - \bar{\varepsilon})^2 \frac{\partial f_0}{\partial \varepsilon}\,\mathrm{d}^3\boldsymbol{k} \tag{14.9}$$

(plus a small correction term if $\boldsymbol{L} \neq \boldsymbol{L}^T$), where $\bar{\varepsilon}$ is a mean energy defined by

$$\int \varepsilon \boldsymbol{v}\cdot\boldsymbol{L}\frac{\partial f_0}{\partial \varepsilon}\,\mathrm{d}^3\boldsymbol{k} = \bar{\varepsilon}\int \boldsymbol{v}\cdot\boldsymbol{L}\frac{\partial f_0}{\partial \varepsilon}\,\mathrm{d}^3\boldsymbol{k}$$

but κ_c is seldom of practical importance. The combination of very large thermoelectric powers with small thermal conductivities makes semiconductors good materials to use for thermoelectric cooling.

If we assume for simplicity that the energy surfaces are spherical, we can easily find the resistance and Hall coefficient of a semiconductor in a magnetic field $\boldsymbol{B}$, by generalizing the results of section 2.4 slightly, in the way we discussed briefly in section 2.6 – we simply have to integrate (2.41) over all energies. Now for $\boldsymbol{B} = 0$, we see from (14.6) and problem 14.2 that the contribution to σ from electrons with ε/kT between u and $u + \delta u$ is

$$\delta\sigma = n_e e^2 \tau_0 u^{3/2+\nu} e^{-u}\delta u \Bigg/ \left[m^* \int_0^\infty u^{3/2} e^{-u}\,\mathrm{d}u\right] \tag{14.10}$$

and from (2.41) it follows that $\sigma_{xx} + i\sigma_{yx}$ is found by multiplying this by $1/(1 - i\omega_c\tau) = 1/(1 - i\omega_c\tau_0 u^\nu)$ and integrating over all u. The integration is easily carried out in the low-field and high-field limits,

$\omega_c\tau_0 \ll 1$ and $\omega_c\tau_0 \gg 1$, and we can then use (11.3) to find the low-field and high-field resistivities ρ_0 and ρ_∞ and the Hall coefficients R_0 and R_∞. We need not write down the full expressions for all these quantities, but simply note that (not surprisingly) $\rho_0 = 1/\sigma$ where σ is given by (14.6), and that (not surprisingly) $R_\infty = -1/n_e e$; this is simply a special case of the very general result (11.11). For phonon scattering, with $v = -\frac{1}{2}$, which is usually the case of interest, we find $\rho_\infty/\rho_0 = 32/9\pi = 1.132$ and $R_0/R_\infty = 3\pi/8 = 1.178$, the values already quoted in section 2.6. For p-type material, of course, $R_\infty = +1/n_h e$. For both p-type and n-type material, the Hall mobility is usually taken to be just $|R_0|/\rho_0$, and these small numerical factors are neglected.

When both electrons and holes are present together, as in intrinsic material, the two-band expressions (11.15) and (11.16) show that we can expect much larger variations of ρ and R_H with the field strength B, and that R_H may change sign as B increases, and these predictions are in good agreement with experiment. Clearly, though, we can no longer write $\mu = |R_H|/\rho$ under these conditions; instead, (11.16) shows that at low fields, $R_H = (\sigma_h\mu_h - \sigma_e\mu_e)/(\sigma_h + \sigma_e)^2$, where $\sigma = 1/\rho$.

14.3 CYCLOTRON RESONANCE AND OPTICAL PROPERTIES

Azbel'–Kaner cyclotron resonance (section 12.1) cannot normally be observed in semiconductors, because the conductivity is so low that the skin depth is large compared with the orbit radius of the carriers, whereas in AKCR experiments it needs to be small. Indeed, at low temperatures a semiconductor may become effectively transparent to microwaves, so that they penetrate it completely. For this to happen, the microwave frequency must exceed the plasma frequency ω_p of (2.35).

This makes possible a different form of cyclotron resonance. If the microwave field is circularly polarized, so that $\boldsymbol{E}$ is rotating at frequency ω about the x axis say, and if a steady field $\boldsymbol{B}$ is applied along x so that the electrons are rotating about x at the cyclotron frequency $\omega_c = eB/m_c^*$, it is intuitively clear that the electrons will absorb power resonantly from the field when $\omega = \omega_c$, because relative to the electrons the field is then not rotating. In fact the rate of power absorption is given by the real part of $\sigma_0 E^2/[1 + i(\omega - \omega_c)\tau]$, where σ_0 is the dc conductivity and τ is some suitable average relaxation time. The power absorbed thus goes through a resonance peak at $B = m_c^* \omega / e$, and the width of the peak is a measure of τ. The cyclotron mass m_c^* is given by (7.7), so that if $\boldsymbol{B}$ lies along the x axis say of the ellipsoid (13.2), $m_c^* = (m_y^* m_z^*)^{1/2}$ (problem 14.3). In a semiconductor such as Si or Ge, with several different conduction electron ellipsoids pointing in different directions, we shall thus see several different resonant peaks, and by varying the direction of $\boldsymbol{B}$ we can determine the orientation of the ellipsoids as well as the effective masses.

The optical properties of semiconductors are broadly similar to those of metals (section 12.2). For frequencies well below the absorption edge, the absorption is small and is described reasonably well by the classical Drude theory. At the absorption edge, electrons begin to be excited from the valence band to the conduction band, and the absorption rises rapidly. As in metals, the joint density of states (12.14) determines the strength of the absorption at and above the edge.

Near the edge, we now need to distinguish between *direct-gap* semiconductors like InSb or GaAs, in which the lowest-energy state in the conduction band and the highest-energy state in the valence band have the same value of $\boldsymbol{k}$ (usually $\boldsymbol{k} = 0$), and *indirect-gap* semiconductors like Ge or Si, in which they do not. In indirect-gap materials, the lowest-energy photon which can produce a vertical transition will need an energy $\hbar\omega_e = \varepsilon_e$, greater than the energy gap ε_g, as shown in Fig. 14.3, because the smallest vertical distance between the bands is ε_e.

It is still possible, however, for photons of energy ε_g to produce *phonon-assisted* transitions, in which momentum is conserved by

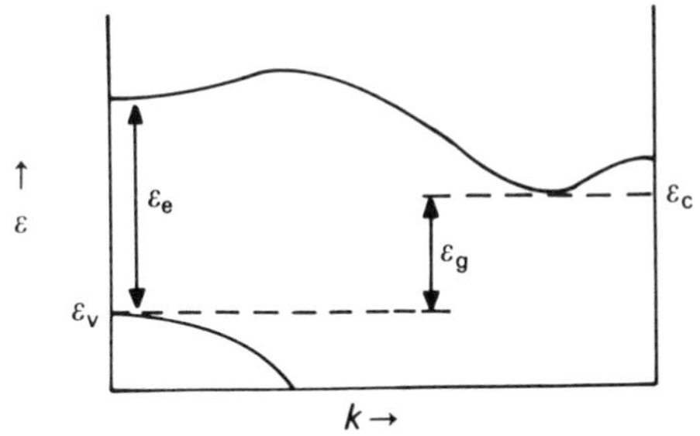

Fig. 14.3 In an indirect-gap semiconductor, a vertical transition needs an energy ε_e greater than ε_g.

emitting or absorbing a phonon of wave-vector $\boldsymbol{q}$ at the same time as the photon is absorbed, so that (12.13) becomes $\boldsymbol{k}' - \boldsymbol{k} = \pm \boldsymbol{q} + \boldsymbol{k}_\phi$. As in metals, $\boldsymbol{k}_\phi$ is negligibly small, and the phonon energy $\hbar\omega_q$ is also small, so that we have $\varepsilon_c(\boldsymbol{k}') - \varepsilon_v(\boldsymbol{k}) = \varepsilon_g = \hbar\omega$ to good approximation, though the phonon energy may produce some fine structure in the spectrum. As one might expect, phonon-assisted transitions have a relatively low transition probability, and for $\hbar\omega > \varepsilon_e$ they are swamped by the much stronger vertical transitions, which do not involve phonons. But in indirect-gap semiconductors, they produce a precursor to the main absorption edge, and enable both ε_e and ε_g to be determined by optical measurements.

In metals, the electrons excited optically from one band to another have no observable effect on the transport properties, because there are too few of them compared with those normally present anyway, but this is not so in semiconductors, where photoconductive effects are readily measurable. In cyclotron resonance experiments, for example, it is necessary to work at low temperatures in order to make τ as long as possible, but this means that the number of thermally excited carriers will be very small. To increase their number and so increase the strength of the cyclotron resonance signal, electrons can be excited optically from the valence band to the conduction band by shining a light on the sample. Likewise in drift mobility experiments, a pulse of light can be used to generate carriers near one end of a bar, which then drift along the bar in the applied electric field.

14.4 ORBIT QUANTIZATION AND THE QUANTUM HALL EFFECT

The de Haas–van Alphen effect and the corresponding resistivity oscillations (section 7.4) cannot normally be observed in semiconductors, because they involve the sudden emptying of quantized orbits as they pass out through the Fermi surface, and normally a semiconductor has no Fermi surface, because ε_F lies in the energy gap. But there are exceptions to this: in InSb and InAs, the donor ionization energy is very small, and falls effectively to zero with moderate doping, because the conduction electrons screen each other from the attractive field of the ionized donors. A finite number of electrons thus remain in the conduction band even at low temperatures, and they then form a degenerate electron gas in which Schubnikov–de Haas resistivity oscillations can be observed.

The quantum Hall effect, whose discovery in 1980 earned von Klitzing the Nobel Prize, also depends on the quantization of orbits in a magnetic field, but this time in an effectively *two-dimensional* conductor. How does one make such a thing? One way is to use a Si MOSFET (metal-oxide-semiconductor field-effect transistor), consisting of an Si wafer, coated on one side with a very thin layer of insulating SiO_2, which is then coated with a layer of metal to form a capacitor with the Si as one plate. If a positive voltage (typically 10–20 V) is applied to the metal plate, a very strong electric field is set up in the SiO_2 layer, which penetrates into the Si crystal to a depth determined by the screening length before it is screened out by the redistribution of charge in the Si. The effect of this field is to lower the potential energy of electrons near the surface relative to those further away, and this can be described in terms of *band bending*: the conduction band edge ε_c falls in energy relative to the Fermi energy ε_F as we approach the surface, and may fall below ε_F at the surface. Because of this, the screening length is much less than (8.10) would predict.

As shown in Fig. 14.4, the band bending produces a very narrow region – typically only 30 Å (3 nm) or so in width – in which ε_F lies within the conduction band, so that a degenerate electron gas can exist. If the z axis is normal to the surface, the electron wave-functions in this region must have the form

$$\psi(r) \sim e^{i\boldsymbol{K}\cdot\boldsymbol{R}} f(z) u_{\boldsymbol{K}}(\boldsymbol{r}) \tag{14.11}$$

where $\boldsymbol{K}$ and $\boldsymbol{R}$ are two-dimensional vectors in the x–y plane and

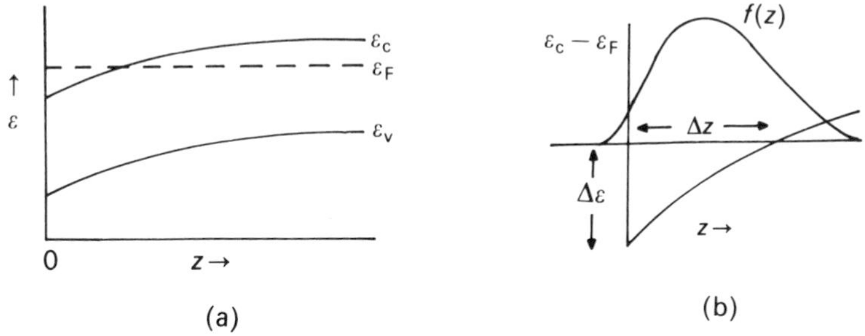

Fig. 14.4 (a) Showing band bending when an electric field is applied to a semiconductor surface. (b) Near the surface, where ε_c is less than ε_F, the lowest allowed states have the form shown.

$f(z)$ falls rapidly to zero outside the region Δz where $\varepsilon_c < \varepsilon_F$. In the lowest-energy solution, $f(z)$ will have the form sketched in Fig. 14.4, with no nodes; other states exist with one or more nodes, but if Δz is small enough these will be of much higher energy, and will not be occupied.

If we now apply a strong magnetic field normal to this degenerate two-dimensional electron gas, the electron orbits in the x–y plane will be quantized in the usual way, and we can think of the states in $\boldsymbol{K}$-space as condensed on to a set of concentric 'Landau rings': in effect, the $k_z = 0$ section through Fig. 7.6. And if, lastly, we pass a current I_x through the system in the x direction, we can measure the resistive voltage V_x and the Hall voltage V_y appearing across the sample in the x and y directions, and hence find the resistivities ρ_{xx} and ρ_{yx}. Because the sample is two-dimensional, ρ_{xx} and ρ_{yx} will now have the dimensions of a resistance. In particular, ρ_{yx} is the ratio of the electric field V_y/Y across the sample (of width Y) to the current per unit width, I_x/Y; in other words, just V_y/I_x, independent of the width Y.

Figure 14.5 shows the remarkable way in which ρ_{xx} and $\rho_{xy}(= -\rho_{yx})$ are found to vary with the magnetic field strength B in such a system. (These data were not obtained on an Si MOSFET, in fact, but on a 'heterostructure' based on GaAs, which behaves very similarly.) The Hall resistance ρ_{xy} rises in steps instead of linearly, and the energy-dissipating component ρ_{xx} is effectively zero except when ρ_{xy}

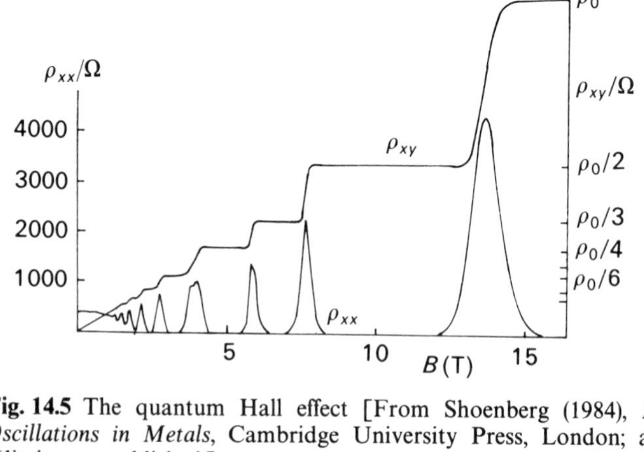

Fig. 14.5 The quantum Hall effect [From Shoenberg (1984), *Magnetic Oscillations in Metals*, Cambridge University Press, London; after von Klitzing, unpublished.]

is changing from one step to the next. Most remarkable of all, each step in ρ_{xy} is given with great accuracy (to better than 1 part in 10^7, in fact) by the expression

$$\rho_{xy} = h/e^2 r = \rho_0/r \quad \text{say,} \tag{14.12}$$

where r is an integer and $\rho_0 = h/e^2 \sim 25\,813\,\Omega$, independent of sample purity and independent of the exact geometry of the sample or of the contacts used to measure V_y. This extraordinary phenomenon thus provides a very precise and convenient resistance standard, and in fact since January 1990 the practical definition of the ohm has been based on the quantum Hall effect, with $\rho_0 = 25\,812.807\,\Omega$.

Why do ρ_{xx} and ρ_{xy} behave in this way? As we shall see, the explanation is quite subtle. Suppose the two-dimensional layer contains n_e electrons per unit area. (In the MOSFET, n_e can be conveniently varied just by altering the 'gate voltage' applied across the capacitor.) Then it is not too difficult to show that (11.12) should still hold in two dimensions, so that $\rho_{xy} = B/n_e e$ as long as $\sigma_{xx} \ll \sigma_{xy}$, where these are the corresponding two-dimensional conductivities. Even if their orbits are quantized, the electrons will still drift at the same rate in the x direction if they are subjected to a field E_y, and this will leave ρ_{xy} unchanged.

In the quantized system, each Landau ring will hold the same number of electrons as would, in $\boldsymbol{B} = 0$, be held in the $\boldsymbol{K}$-space area $\Delta\mathscr{A}$ between one ring and the next. For each spin orientation this is $\Delta\mathscr{A}/4\pi^2$ per unit area of sample, by the two-dimensional equivalent of (1.23). Using $\Delta\mathscr{A} = 2\pi eB/\hbar$, from (7.14), this becomes simply eB/h. Thus whenever

$$eB/h = n_e/r \tag{14.13}$$

exactly r Landau levels will be completely filled, and all higher levels will be empty. (At the low temperatures and high fields at which these experiments are carried out, kT is extremely small compared with $\hbar\omega_c$, so that there is negligible thermal excitation from one Landau level to another.) So if $\rho_{xy} = B/n_e e$, this means that whenever $\rho_{xy} = h/e^2 r$ – the values at which the steps appear – exactly r Landau levels are filled. Moreover, whenever exactly r levels are filled, the conductivity component σ_{xx} will vanish, because there is then an energy gap of $\hbar\omega_c$ between the filled states in the rth level and the nearest unfilled states in the $(r+1)$th level. Because of this energy gap, the electrons cannot gain energy from an applied field, or lose energy by collisions, so that $\sigma_{xx} = 0$. If σ_{xx} vanishes while σ_{xy} remains finite, (11.3) shows that ρ_{xx} will also vanish.

This leads to what looks like a rather simple explanation of the quantum Hall effect: except when one of the Landau levels lies exactly at ε_F, we *expect* $\rho_{xy} = h/e^2r$ and $\rho_{xx} = 0$, with r equal to the number of filled levels below ε_F. On this picture, the energy of the highest (rth) filled level will rise towards ε_F as B increases, but as long as it stays below ε_F and remains filled we shall have $n_e = eBr/h$, so that $\rho_{xy} = h/e^2r$. When the rth level reaches ε_F, it will begin to empty, and ρ_{xy} will begin to rise; at the same time, ρ_{xx} will become non-zero because there are now unfilled states close in energy to the filled states. Once the rth level has completely emptied, ρ_{xy} will take on the constant value $h/e^2(r-1)$, and ρ_{xx} will revert to zero, precisely as observed experimentally.

Unfortunately, this explanation cannot be right as it stands, because if $\rho_{xy} = B/n_e e$, the stepwise variation of ρ_{xy} as B varies must be accompanied by a sawtooth variation of n_e, whereas simple electrostatics shows that n_e cannot vary in this way. The charge density $n_e e$ in the surface layer has to screen the semiconductor from the field E (normal to the surface) produced by the applied gate voltage, and it at once follows that $n_e = \varepsilon_r \varepsilon_0 E/e$, where ε_r is the dielectric constant of the SiO_2 layer. But if n_e remains constant, and if $\rho_{xy} = B/n_e e$, there can be no quantized steps in ρ_{xy}. The fallacy in our argument was to assume that the uppermost filled Landau level almost always lay *below* ε_F, so that it was completely full. In fact, if n_e is to remain constant, the uppermost Landau level must always be only partly filled, except when (14.3) is exactly satisfied, and must therefore lie exactly at ε_F (problems 14.4, 14.5).

To explain the quantum Hall effect, we must find a model in which ε_F can lie *between* Landau levels. To do this, we suppose that the crystal structure in the surface layer is sufficiently disordered to produce substantial broadening of the Landau levels, and we further suppose that within each broadened level, only the states near the centre of the level are extended current-carrying states of the form (14.11). Away from the centre of the level, the states are *localized*, as shown in Fig. 14.6. (This kind of localization is discussed briefly in section 15.3.) The localized states then act as a kind of reservoir, so that the number n_{ex} of electrons in *extended* states can vary with B, even though the total number n_e cannot. As long as ε_F lies outside an extended-state (e-s) region – say between the rth and $(r+1)$th

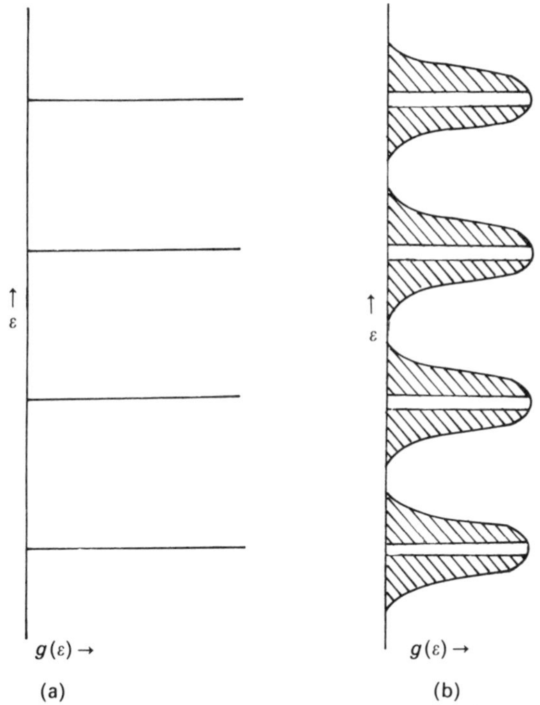

Fig. 14.6 (a) The sharp Landau levels in a good crystal. (b) The broadened levels in a disordered crystal. The disorder produces localization of the states in the wings of each broadened level, as shown shaded; only the unshaded states near the centre of the level remain extended and capable of carrying a current.

regions – exactly r sets of e-s levels will be filled, and n_{ex} will grow linearly with B (while the number of localized states, $e_e - n_{ex}$, will correspondingly fall). As B increases further, the rth e-s region will pass through ε_F, and n_{ex} will fall as this set of states empties. Thus n_{ex} now varies in sawtooth fashion, even though n_e if fixed.

Now we can show, surprisingly, that if r sets of e-s levels are filled (and all higher e-s levels are empty) the Hall resistance ρ_{xy} is simply ρ_0/r still, unaffected by the broadening of the Landau levels or by the existence of localized states. To see why, consider Fig. 14.7. The shaded areas here represent localized states, so that in the unshaded areas only extended states exist. In these unshaded areas, the density of states must therefore be the same as in the unbroadened

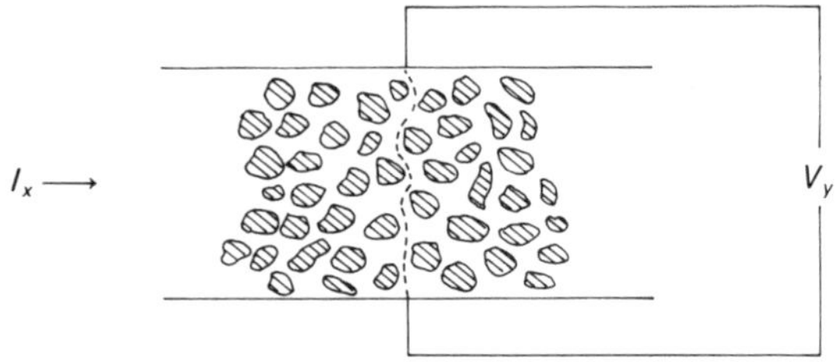

Fig. 14.7 The localized states are localized in the regions in the surface shown shaded. In the unshaded regions only extended states exist, and the measured Hall voltage V_y depends only on the field E_y in the unshaded regions.

case: eB/h states per Landau level, per unit area. If r sets of e-s levels are completely filled, the number of e-s electrons per unit area in the unshaded regions will thus be $n_{ex} = eBr/h$. It follows that the Hall field E_y must be the same as if the levels were unbroadened: $E_y = J_x B/n_{ex}e = J_x h/e^2 r$. To find the Hall voltage V_y, we simply have to integrate E_y over the width Y of the sample, taking care that the path of integration passes only through unshaded regions. We thus find

$$V_y = \int_0^Y E_y \, dy = (h/e^2 r) \int_0^Y J_x \, dy = (\rho_0/r) I_x \qquad (14.14)$$

In other words, we find $\rho_{xy} = \rho_0/r$ whenever ε_F lies between extended-state regions, and the steps in the quantum Hall effect are thus accounted for. The breadth of the steps depends on the relative numbers of extended states and localized states (problem 14.6).

Finally, there is yet one more twist to the story. A year or two after von Klitzing's original discovery, it was found that in carefully prepared samples, steps in ρ_{xy} appeared for *fractional* values of r in (14.12) – not for all fractions, but for fractions such as 1/3, 2/3, 4/3, 5/3; 2/5, 3/5, 4/5; 2/7,... This fractional quantum Hall effect proved a great deal more difficult to explain theoretically than the integral version, and the explanation is a very subtle one, in which electron–electron interactions play an essential part. In effect, these interactions give rise to excitations which behaves like particles of fractional charge, $e/3$, $e/5$, etc. and these in turn give rise to the fractional values of r. But the details of the theory are far beyond the scope of this book.

15

Other topics

15.1 CHARGE DENSITY WAVES

There are many topics which we have had to leave untouched in this book, simply to keep it of reasonable length. We have said nothing, for example, about semiconductor devices – p–n junctions, transistors, light-emitting diodes and so on – and nothing about amorphous and liquid metals and semiconductors, where the lack of a regular crystal structure means that Bloch waves and Brillouin zones and well-defined energy bands no longer exist. We have said very little about superconductivity, and nothing about magnetically ordered materials like the ferromagnetic metals and the rare earth metals, in which the electron spins line up cooperatively in various ways. To deal with any of these subjects properly would take another book.

There are, however, a few other topics which we *shall* discuss briefly in this final chapter, because they follow on naturally from the things that we have talked about earlier, and because they are currently of considerable research interest. We begin with charge density waves.

These occur because the energy of the conduction electrons in metallic materials can sometimes be lowered significantly by a periodic distortion of the crystal structure. To see how, suppose for example that the distortion causes the length of the unit cell in the x direction, a_1 say (cf. (3.1)), to vary periodically with x, so that $a_1 = a_{01}(1 + b\cos k_c x)$. We assume here that $b \ll 1$ and $k_c \ll 2\pi/a_{01}$, so that the scale of the distortion is small, and its wavelength is large, compared with the unit cell size.

This distortion will give rise to components in the potential $V(\boldsymbol{r})$ with wave-vectors $\pm k_c$, and these will then couple together the

Bloch states at $\boldsymbol{k}$ and $\boldsymbol{k} \pm \boldsymbol{k}_c$, just as in (4.19). As there, the coupling will be particularly strong if the states at $\boldsymbol{k}$ and $\boldsymbol{k} \pm \boldsymbol{k}_c$ have similar energies, and it will then perturb them apart, just as the states of Fig. 4.2 have been perturbed apart from those of Fig. 4.1. Again, as there, the result will be to *lower the energy* of the states below the gap. We can now see how a charge density wave can arise: if the undistorted material contains two sizeable areas of Fermi surface which are a distance $\boldsymbol{k}_c$ apart in $\boldsymbol{k}$-space, then a CDW can lower their energy significantly by coupling them in this way. If the resultant reduction of energy exceeds the elastic energy needed to produce the distortion, it will then occur spontaneously, at least at low temperatures.

A surprising number of materials in fact exhibit CDW formation at low temperatures, though none of the simple metals do. (It has from time to time been suggested that potassium does, but there is little evidence for this so far.) Of the metallic compounds which develop CDWs, $NbSe_3$ has been studied particularly closely because of the interesting transport properties that the movement of the CDWs gives rise to.

15.2 DISORDERED ALLOYS

Some alloys, such as CuZn, can be prepared in either an ordered or a disordered state. In the ordered state, the alloy still has a regular periodic structure, with alternate lattice sites occupied by Cu and Zn atoms, so that the band structure can still be calculated using the techniques of section 5.1. But most alloys (including CuZn unless it is cooled very slowly) are disordered – the atoms of the constituent metals are distributed at random on the available lattice sites. It is not then obvious how to calculate the band structure, or indeed whether a band structure still exists. The existence of Bloch function solutions of the Schrödinger equation depends on the fact that the potential $V(\boldsymbol{r})$ is periodic, and in a disordered alloy it no longer is. We saw in section 8.1 how to find the rate at which electrons were scattered out of a given Bloch state $\boldsymbol{k}$ when there was just one impurity present, but now the scattering is so strong that a state of given $\boldsymbol{k}$ no longer has a well-defined energy; because the lifetime of the electron in such a state is so short, the energy is correspondingly broadened, by the uncertainty principle.

Methods have been developed – notably the 'coherent potential approximation' – for extending the techniques of section 5.1 to deal with this problem, and to calculate the broadened and complex energy spectra of electron states in disordered alloys. The results can then be compared with the results of ARPES experiments (section 12.2) and this combination of theory and experiment is leading to a much better understanding of the properties of alloys.

15.3 LOCALIZATION

Even in a perfectly periodic structure, it is not always true that the relevant wave-functions are Bloch waves. Consider for example a semiconductor containing N_d donor atoms, and suppose (unrealistically) that these donors are spaced at regular intervals throughout the crystal, so that they themselves lie on a large-scale 'lattice'. Then, if $kT \ll \varepsilon_d$, where ε_d is the donor ionization energy, the conduction band will be empty and each donor state will contain one electron. Now, if the wave-functions of the electrons on adjacent donor atoms overlap appreciably, the approach of section 5.1 would lead us to expect these electrons to form a half-filled band, producing a reasonably good conductor. But this would be to overlook the importance of the correlation forces discussed in section 6.1, which are ignored in the simple one-electron approximation of 5.1. Because of these Coulomb repulsive forces between the electrons, it would be energetically very unfavourable to have more than one electron on each donor atom, and the electrons are in fact effectively localized rather than mobile. Only if the electron density is high enough for screening to become effective, in the way discussed below eq. (6.15), will the electrons become mobile. And when the screening *is* effective, the electrons may be mobile even if the donors are not arranged on a regular lattice.

We thus expect, at low temperatures, a transition from a non-conducting state to a conducting state when N_d exceeds some critical value, and this 'Mott transition', predicted by Mott in 1949, is indeed observed, as remarked briefly in the first paragraph of section 14.4. Below the transition, the material may in fact still conduct, by a different mechanism, if it is partially compensated – that is, if there are N_a acceptors present as well as the N_d donors, with $N_a < N_d$. At low temperatures, we shall then have N_a ionized donors, which have lost their electrons to the acceptors, and $N_d - N_a$ neutral donors, still holding electrons. These electrons can then move through the crystal by 'hopping' on to an ionized donor, neutralizing it and leaving an ionized one behind. The hopping process proceeds by thermally activated tunnelling, with an activation energy ε_A, and leads to a conductivity varying as $\exp(-\varepsilon_A/kT)$ (or in some situations, where electrons have a choice of sites to hop to with different values of ε_A, through 'variable-range hopping' varying as $\exp(-C/T^{1/4})$).

A different kind of localization, not dependent on correlation effects, was predicted in 1958 by Anderson (who shared the Nobel Prize with Mott in 1977). He showed that an electron moving through an array of potential wells would become localized if the wells, instead of all being of the same depth, fluctuated randomly in depth by more than a certain amount. The states near the top and bottom of the band are the first to become localized as the depth fluctuations become larger, while those near the centre stay non-localized, and this leads to the idea of a 'mobility edge' in energy between the localized and non-localized states.

Since about 1975, there has been a great deal of work on localization and disorder, both experimental and theoretical. Rather surprisingly, it appears that these two physically quite different mechanisms – electron–electron interactions on the one hand and disorder on the other – often lead to rather similar behaviour, and it is not always easy to distinguish between them. In recent years there has been an increasing amount of work on *low-dimensional* systems – thin films or surface layers (as in the quantum Hall effect) and thin wires – because the effects of interactions and of localization are predicted to be significantly different in low-dimensional systems. As we saw in section 14.4, Anderson localization plays an important part in the explanation of the quantum Hall effect.

APPENDIX A

The semi-classical Hamiltonian and the Schrödinger equation in a magnetic field

In this appendix, we look briefly at the Hamiltonian form of the equations of motion of classical mechanics, in order to show that he behaviour of a Bloch electron can be described in terms of a semi-classical Hamiltonian, and in order to derive the form of the Schrödinger equation for a particle in a magnetic field.

The energy of a classical particle, when written in Hamiltonian form, is a function of $\boldsymbol{r}$ and of a momentum variable $\boldsymbol{p}$, 'conjugate' to $\boldsymbol{r}$: $\varepsilon = \varepsilon_H(\boldsymbol{p}, \boldsymbol{r})$, where the subscript H shows that ε_H is written in Hamiltonian form. (More generally, $\boldsymbol{r}$ may be replaced by a 'generalized' position variable $\boldsymbol{q}$, and one often sees (7.11) written in terms of $\oint \boldsymbol{p}\cdot d\boldsymbol{q}$ rather than $\oint \boldsymbol{p}\cdot d\boldsymbol{r}$, but we shall not need this degree of generality.) The momentum $\boldsymbol{p}$ is said to be conjugate to $\boldsymbol{r}$ if the equations

$$\dot{\boldsymbol{r}} = \partial\varepsilon_H/\partial\boldsymbol{p}; \quad \dot{\boldsymbol{p}} = -\partial\varepsilon_H/\partial\boldsymbol{r} \qquad \text{(A.1, A.2)}$$

are satisfied; these two equations are the equations of motion of the particle in Hamiltonian form.

Thus, for a particle of charge $-e$ in a uniform field $\boldsymbol{E}$ (with $\boldsymbol{B} = 0$), we have $\boldsymbol{p} = m\boldsymbol{v}$, and

$$p^2/2m + e\boldsymbol{E}\cdot\boldsymbol{r} = \varepsilon_H \qquad \text{(A.3)}$$

Then (A.1) correctly yields $\dot{\boldsymbol{r}} = \boldsymbol{p}/m = \boldsymbol{v}$, and (A.2) correctly yields $\dot{\boldsymbol{p}} = m\dot{\boldsymbol{v}} = -e\boldsymbol{E}$. But if $\boldsymbol{B} \neq 0$, (A.3) needs modifying, if (A.2) is to lead correctly to the Lorentz force equation. In fact we now need to define the 'momentum' $\boldsymbol{p}$ not as $m\boldsymbol{v}$ but as $m\boldsymbol{v} - e\boldsymbol{A}$, where $\boldsymbol{A}(\boldsymbol{r})$ is the vector potential describing the magnetic field, so that curl $\boldsymbol{A} = \boldsymbol{B}$.

The expression for ε_H then becomes

$$(\boldsymbol{p} + e\boldsymbol{A})^2/2m + e\boldsymbol{E}\cdot\boldsymbol{r} = \varepsilon_H \tag{A.4}$$

where the first term is still just the kinetic energy $\frac{1}{2}mv^2$, but expressed in unfamiliar form.

Applied to (A.4), (A.1) correctly yields $\dot{\boldsymbol{r}} = (\boldsymbol{p} + e\boldsymbol{A})/m = \boldsymbol{v}$, and (A.2) correctly yields $m\dot{\boldsymbol{v}} = -e(\boldsymbol{E} + \boldsymbol{v} \times \boldsymbol{B})$, though it takes a little care to show this. (In evaluating $-\partial\varepsilon_H/\partial\boldsymbol{r}$, we have to remember that $\boldsymbol{A} = \boldsymbol{A}(\boldsymbol{r})$, so that, for example,

$$-\partial\varepsilon_H/\partial x = -eE_x - (e/m)(\boldsymbol{p} + e\boldsymbol{A})\cdot\partial\boldsymbol{A}/\partial x = -eE_x - e\boldsymbol{v}\cdot\partial\boldsymbol{A}/\partial x$$

and, in interpreting $\dot{\boldsymbol{p}}$, we have to remember that if the electron is moving through a non-uniform potential $\boldsymbol{A}(\boldsymbol{r})$, the value of $\boldsymbol{A}$ 'seen' by the electron will vary with time, so that, for example,

$$\dot{p}_x = m\dot{v}_x - e\boldsymbol{v}\cdot(\partial A_x/\partial\boldsymbol{r}).)$$

Now (A.3) and (A.4) apply to a classical point particle, and certainly do not apply to a Bloch electron moving through a crystal lattice. For such an electron we have $\varepsilon = \varepsilon(\boldsymbol{k})$, if there is no applied field $\boldsymbol{E}$ or $\boldsymbol{B}$. If there is an applied field $\boldsymbol{E}$, and if the electron is thought of as a wave-packet localized near the point $\boldsymbol{r}$, we can write the total energy as $\varepsilon(\boldsymbol{k}) + e\boldsymbol{E}\cdot\boldsymbol{r}$. Now, although we stressed in section 7.1 that $\hbar\boldsymbol{k}$ was not the true momentum of the electron, it turns out that if we treat $\hbar\boldsymbol{k}$ as the momentum variable conjugate to $\boldsymbol{r}$, the resultant 'semi-classical Hamiltonian'

$$\varepsilon(\boldsymbol{p}) + e\boldsymbol{E}\cdot\boldsymbol{r} = \varepsilon_H \tag{A.5}$$

with $\boldsymbol{p} = \hbar\boldsymbol{k}$, correctly describes the motion of the electron. Thus (A.1) now yields precisely the result (7.1), and (A.2) yields the result $\hbar\dot{\boldsymbol{k}} = -e\boldsymbol{E}$. Moreover, if $\boldsymbol{B} \neq 0$, (A.1) and (A.2) continue to give the correct results if we replace (A.5) by

$$\varepsilon(\boldsymbol{p} + e\boldsymbol{A}) + e\boldsymbol{E}\cdot\boldsymbol{r} = \varepsilon_H \tag{A.6}$$

that is, if we write $\hbar\boldsymbol{k} = \boldsymbol{p} + e\boldsymbol{A}$. Then (A.1) again yields (7.1), and (A.2) correctly yields $\hbar\dot{\boldsymbol{k}} = -e(\boldsymbol{E} + \boldsymbol{v} \times \boldsymbol{B})$.

On this basis, then, the semi-classical momentum variable for a Bloch electron in a magnetic field is $\boldsymbol{p} = \hbar\boldsymbol{k} - e\boldsymbol{A}$, and this is the expression we used in (7.12).

We can derive the Schrödinger equation for a particle from the classical Hamiltonian by replacing $\boldsymbol{p}$ by the operator $-i\hbar\nabla$. Thus, if the classical Hamiltonian is

$$\boldsymbol{p}^2/2m + V(\boldsymbol{r}) = \varepsilon_{\mathrm{H}} \tag{A.7}$$

the corresponding Schrödinger equation has the familiar form

$$-(\hbar^2/2m)\nabla^2\psi(\boldsymbol{r}) + V(\boldsymbol{r})\psi(\boldsymbol{r}) = \varepsilon\psi(\boldsymbol{r})$$

(cf. (4.1)). The Schrödinger equation for a particle in a magnetic field may be less familiar: the classical Hamiltonian then has the form

$$(\boldsymbol{p} + e\boldsymbol{A})^2/2m + V(\boldsymbol{r}) = \varepsilon_{\mathrm{H}}$$

and it follows that the Schrödinger equation is

$$(-i\hbar\nabla + e\boldsymbol{A})^2\psi(\boldsymbol{r})/2m + V(\boldsymbol{r})\psi(\boldsymbol{r}) = \varepsilon\psi(\boldsymbol{r}). \tag{A.8}$$

For $V(\boldsymbol{r}) = 0$, this is the equation to a free electron in a magnetic field, which is not too difficult to solve; the allowed energies are then given by (7.10).

APPENDIX B

The Boltzmann equation and the collision integral

Suppose that in volume element $\delta^3\boldsymbol{k}\,\delta^3\boldsymbol{r}$ there are $f(\boldsymbol{k},\boldsymbol{r})\delta^3\boldsymbol{k}\,\delta^3\boldsymbol{r}/4\pi^3$ electrons. These electrons will be moving through real space at velocity $\dot{\boldsymbol{r}} = \boldsymbol{v}$ and through $\boldsymbol{k}$-space at velocity $\dot{\boldsymbol{k}} = \boldsymbol{F}/\hbar$, where $\boldsymbol{F}$ is the force produced by applied fields. The Boltzmann equation is derived by arguing that in the absence of collisions, the electrons at point $\boldsymbol{k}, \boldsymbol{r}$ at time t will be the electrons which were at $\boldsymbol{k} - \delta\boldsymbol{k}$, $\boldsymbol{r} - \delta\boldsymbol{r}$ at time $t - \delta t$, where $\delta\boldsymbol{k} = \dot{\boldsymbol{k}}\,\delta t, \delta\boldsymbol{r} = \dot{\boldsymbol{r}}\,\delta t$:

$$\begin{aligned} f(\boldsymbol{k},\boldsymbol{r},t) &= f(\boldsymbol{k} - \delta\boldsymbol{k}, \boldsymbol{r} - \delta\boldsymbol{r}, t - \delta t) \\ &= f(\boldsymbol{k},\boldsymbol{r},t) - \left(\dot{\boldsymbol{k}}\cdot\frac{\partial f}{\partial \boldsymbol{k}} + \dot{\boldsymbol{r}}\cdot\frac{\partial f}{\partial \boldsymbol{r}} + \frac{\partial f}{\partial t}\right)\delta t \end{aligned}$$

(We have assumed here that the size of the volume element $\delta^3\boldsymbol{k}\,\delta^3\boldsymbol{r}$ occupied by the electrons does not change, and this can indeed be shown to be true.) Thus, if there are no collisions, the rate of change of $f(\boldsymbol{k},\boldsymbol{r})$ is given by

$$\frac{\partial f}{\partial t} = -\dot{\boldsymbol{k}}\cdot\frac{\partial f}{\partial \boldsymbol{k}} - \dot{\boldsymbol{r}}\cdot\frac{\partial f}{\partial \boldsymbol{r}}$$

Adding a collision term, we obtain the Boltzmann equation:

$$\frac{\partial f}{\partial t} = -\dot{\boldsymbol{k}}\cdot\frac{\partial f}{\partial \boldsymbol{k}} - \dot{\boldsymbol{r}}\cdot\frac{\partial f}{\partial \boldsymbol{r}} + \left[\frac{\partial f}{\partial t}\right]_{\text{coll.}} \tag{B.1}$$

If we write $f = f_0 + f_1$, where f_0 is the FD function (1.11), we can replace f by f_1 in the two $\partial f/\partial t$ terms, since by definition $\partial f_0/\partial t = 0$, and by f_0 in the remaining terms, since $\partial f_1/\partial \boldsymbol{k}$ and $\partial f_1/\partial \boldsymbol{r}$ are usually negligibly small. As an exception to this, if a magnetic field is present, $\dot{\boldsymbol{k}}$ will include a term $\dot{\boldsymbol{k}}_\mathrm{B} = -e\boldsymbol{v} \times \boldsymbol{B}/\hbar$, and we must then retain the term $\dot{\boldsymbol{k}}_\mathrm{B} \cdot \partial f_1/\partial \boldsymbol{k}$, because $\dot{\boldsymbol{k}}_\mathrm{B} \cdot \partial f_0/\partial \boldsymbol{k}$ vanishes. This means that the Boltzmann equation becomes much more difficult to solve when $\boldsymbol{B} \neq 0$.

Unless the applied fields are time-dependent, we shall be looking for steady-state solutions of (B.1) in which $\partial f/\partial t = 0$, so that the collision term exactly balances the 'driving' terms in $\dot{\boldsymbol{k}}$ and $\dot{\boldsymbol{r}}$. For the collision term, we assume as in Chapter 8 that the probability of scattering from $\boldsymbol{k}$ to $\delta^3\boldsymbol{k}'$ is given by $P_{\boldsymbol{kk}'}\delta^3\boldsymbol{k}'$, but we now take into account the fact – tacitly ignored in Chapter 8 – that scattering can only take place if state $\boldsymbol{k}'$ is empty, because of the exclusion principle. The probability that $\boldsymbol{k}$ is occupied is $f_{\boldsymbol{k}}$, and the probability that $\boldsymbol{k}'$ is empty is $1 - f_{\boldsymbol{k}'}$, so that $f_{\boldsymbol{k}}$ changes, because of outward scattering from $\boldsymbol{k}$ into all other states $\boldsymbol{k}'$, at a rate

$$[\partial f_{\boldsymbol{k}}/\partial t]_{\mathrm{out}} = -\int \mathrm{d}^3\boldsymbol{k}' P_{\boldsymbol{kk}'} f_{\boldsymbol{k}}(1 - f_{\boldsymbol{k}'}) \tag{B.2}$$

To balance this, electrons are at the same time being scattered *into* state $\boldsymbol{k}$ from all other states, at a rate

$$[\partial f_{\boldsymbol{k}}/\partial t]_{\mathrm{in}} = \int \mathrm{d}^3\boldsymbol{k}' P_{\boldsymbol{k}'\boldsymbol{k}} f_{\boldsymbol{k}'}(1 - f_{\boldsymbol{k}}) \tag{B.3}$$

so that the net rate of change of $f_{\boldsymbol{k}}$ due to scattering, both in and out, is

$$[\partial f_{\boldsymbol{k}}/\partial t]_{\mathrm{coll.}} = -\int \mathrm{d}^3\boldsymbol{k}' [P_{\boldsymbol{kk}'} f_{\boldsymbol{k}}(1 - f_{\boldsymbol{k}'}) - P_{\boldsymbol{k}'\boldsymbol{k}} f_{\boldsymbol{k}'}(1 - f_{\boldsymbol{k}})] \tag{B.4}$$

Now in thermal equilibrium, when $f_{\boldsymbol{k}} = f_{0,\boldsymbol{k}}$ and $f_{\boldsymbol{k}'} = f_{0,\boldsymbol{k}'}$, this must vanish, so that

$$P_{\boldsymbol{kk}'} f_{0,\boldsymbol{k}}(1 - f_{0,\boldsymbol{k}'}) = P_{\boldsymbol{k}'\boldsymbol{k}} f_{0,\boldsymbol{k}'}(1 - f_{0,\boldsymbol{k}}) \tag{B.5}$$

We assume that $P_{\boldsymbol{kk}'}$ is unaffected by the applied fields; this is true for defect scattering, and almost always a good approximation for phonon scattering (though 'phonon-drag' effects may cause $n_{\boldsymbol{q}}$ to depart slightly from its equilibrium value (8.12)). We can then use (B.5) to simplify (B.4) considerably. Consider first elastic scattering, so that $P_{kk'}$ and $P_{k'k}$ vanish unless $\varepsilon_{k'} = \varepsilon_k$. We can therefore put $f_{0,k'} = f_{0,k}$ in (B.5), which then reduces to $P_{kk'} = P_{k'k}$ – for elastic scattering, these two coefficients must be equal. And in that case, (B.4) reduces to the much simpler form

$$[\partial f_k/\partial t]_{\text{coll.}} = -\int \mathrm{d}^3\boldsymbol{k}'\, P_{kk'}(f_k - f_{k'}) \tag{B.6}$$

The exclusion principle factors $(1 - f_{k'})$ and $(1 - f_k)$ in (B.4) have vanished, essentially because they affect inward and outward scattering by exactly the same amount. The effective outward scattering rate becomes $-\int \mathrm{d}^3\boldsymbol{k}'\, P_{kk'} f_k$, $= -f_k/\tau_k^0$; precisely what we took it to be in Chapter 8.

We saw in Chapter 8 that the effective scattering rate $1/\tau_k^e$ might be much less than $1/\tau_k^0$ unless each collision was fully randomizing, and that transport problems could be simplified by making the relaxation time approximation, which assumes that each collision *is* fully randomizing. In terms of the Boltzmann equation, the corresponding approximation involves replacing $f_{k'}$ by $f_{0,k}$ in (B.6), which then takes the simple form

$$[\partial f_{1,k}/\partial t]_{\text{coll.}} = -f_{1,k}/\tau_k^0 \tag{B.7}$$

where $f_{1,k} = f_k - f_{0,k}$. In replacing $f_{k'}$ by $f_{0,k}$, we are assuming that

$$\begin{aligned}\int \mathrm{d}^3\boldsymbol{k}'\, P_{kk'} f_{k'} &= \int \mathrm{d}^3\boldsymbol{k}'\, P_{kk'} f_{0,k} \\ &= \int \mathrm{d}^3\boldsymbol{k}'\, P_{kk'} f_{0,k'}\end{aligned} \tag{B.8}$$

since $f_{0,k} = f_{0,k'}$. To see what this implies, note that we must in any case have $\int \mathrm{d}^3\boldsymbol{k}'\, f_{k'} = \int \mathrm{d}^3\boldsymbol{k}'\, f_{0,k'}$, if the conductor is to remain uncharged. It follows that (B.8) should be valid if $P_{kk'}$ is independent of the direction of $\boldsymbol{k}'$. In other words, the relaxation time approximation in the form (B.7) involves precisely the same assumption, of fully randomizing collisions, that we made in Chapter 8.

For inelastic scattering, we can still use (B.5) to simplify (B.4), though we no longer have $f_{0,k'} = f_{0,k}$. If we write $f_k = f_{0,k} + f_{1,k}$, $f_{k'} = f_{0,k'} + f_{1,k'}$ and neglect product terms $f_{1,k} f_{1,k'}$ (which would lead to departures from Ohm's law), we find that (B.4) reduces to

$$[\partial f_k/\partial t]_{\text{coll.}} = -\int d^3k'[P_{kk'}(h_{k'}/h_k)f_{1,k} - P_{k'k}(h_k/h_{k'})f_{1,k'}] \quad \text{(B.9)}$$

where $h_k = 1 - f_{0,k}$, $h_{k'} = 1 - f_{0,k'}$. Writing $f_{1,k} = f_k - f_{0,k}$, we see that the effective outward scattering rate – the term proportional to f_k – is now

$$[\partial f_k/\partial t]_{\text{out}} = -f_k/\tau_k^0$$

with

$$1/\tau_k^0 = \int d^3k' P_{kk'}(h_{k'}/h_k) \quad \text{(B.10)}$$

whereas (8.20) and (8.25) omitted the $(h_{k'}/h_k)$ factor. As it happens, the resultant error is largely compensated if we also replace the factor $n_q + 1$ in (8.15) by n_q, as we did in (8.19). To correct for both errors, (8.19) should be multiplied by the factor

$$[(n_q + 1)h_{k'}^- + n_q h_{k'}^+]/2n_q h_k \quad \text{(B.11)}$$

where $h_{k'}^- = 1 - f_0(\varepsilon_k - \hbar\omega_q)$ and $h_{k'}^+ = 1 - f_0(\varepsilon + \hbar\omega_q)$. Over the relevant range of values of $\hbar\omega_q/kT$ and $(\varepsilon_k - \varepsilon_F)/kT$, this factor approximates to $1 + \hbar\omega_q/2kT$ – near enough to 1 for our purposes.

The first and second terms in (B.11) represent scattering in which the electron loses energy $\hbar\omega_q$ and gains energy $\hbar\omega_q$ respectively. In the approximation (8.19) the two processes are equally likely, which would mean that if an electron started with energy $\varepsilon_k > \varepsilon_F$, phonon scattering would do nothing to reduce its average energy to ε_F. But the correction factor (B.11) shows that the two processes are not equally likely; in fact, the first term in (B.11) ($S_{kk'}^-$, say) is always larger than the second ($S_{kk'}^+$) if $\varepsilon_k > \varepsilon_F$, and smaller if $\varepsilon_k < \varepsilon_F$, so that the average energy of the electron always tends back to ε_F. If we write $\eta_k = (\varepsilon_k - \varepsilon_F)/kT$ and $\eta_{k'}^{\pm} = (\varepsilon_k - \varepsilon_F \pm \hbar\omega_q)/kT$, the average value of η after collision is given by

$$\bar{\eta}_{k'} = \frac{(S_{kk'}^+\eta_{k'}^+ + S_{kk'}^-\eta_{k'}^-)}{(S_{kk'}^+ + S_{kk'}^-)},$$

from which we find, to good approximation,

$$\bar{\eta}_{k'}/\eta_k = 1 - (\hbar\omega_q/2kT)^2 \tag{B.12}$$

for $\hbar\omega_q \gtrsim kT$; $\bar{\eta}_{k'}/\eta_k \to 0$ for $\hbar\omega_q \gtrsim kT$. Now the expression for the thermal vector mean free path $\boldsymbol{L}_k^T$, corresponding to (8.23) for $\boldsymbol{L}_k$, is

$$\eta_k \boldsymbol{L}_k^T = \eta_k \boldsymbol{v}_k \tau_k^0 + \tau_k^0 \int \mathrm{d}^3\boldsymbol{k}'\, P_{kk'} \eta_{k'} \boldsymbol{L}_{k'}^T \tag{B.13}$$

and the approximation which led from (8.23) to (8.25) now leads to

$$1/\tau_k^T = \int \mathrm{d}^3\boldsymbol{k}' P_{kk'}[1 - (\bar{\eta}_{k'}/\eta_k)\cos\theta]. \tag{B.14}$$

At high temperatures, where $\hbar\omega_{\max} < kT$ and where scattering is mainly through large angles, the resultant value of $1/\tau_k^T$ does not differ significantly from $1/\tau_k^0$. But for low-temperature phonon scattering, the replacement of $1-\cos\theta$ by $1-(\bar{\eta}_{k'}/\eta_k)\cos\theta$ effectively removes a T^2 factor from the scattering rate, so that $1/\tau_k^T$ varies as T^3 instead of T^5.

Problems

(Relevant fundamental constants are given on p. 223)

1 The free electron model

1.1 Calculate the rms thermal velocity v_r from (1.1) for a classical electron gas at 300 K.

1.2 Cu (AW = 63.5) has a density of $8.89 \times 10^3\,\mathrm{kg\,m^{-3}}$ and an electrical conductivity of $5.8 \times 10^7\,\mathrm{ohm^{-1}\,m^{-1}}$ at 300 K. Assuming one conduction electron per atom, use (1.7) to work out the Drude value of l. If the thermal conductivity at 300 K is $385\,\mathrm{W\,m^{-1}\,K^{-1}}$, use (1.9) to work out a second estimate of l.

1.3 Use of a table of physical constants to find values of σ (or ρ) and κ for several metals, at several temperatures if possible, and work out the values of the Lorenz number $\mathscr{L}$.

1.4 Show that the numerical coefficient of the k^2T^2 term in (1.15) is equal to $2\int_0^\infty x\,\mathrm{d}x/(e^x + 1)$, and that this is equal to $2(1 - 1/4 + 1/9 - 1/16\ldots)$ – which in fact sums to $\pi^2/6$.

1.5 Work out the Fermi energy $\varepsilon_{\mathrm{F},0}$ the Fermi wave-number $k_{\mathrm{F},0}$ and the Fermi velocity $v_{\mathrm{F},0}$ for Cu on the free-electron model, using the data of problem 1.2.

2 Properties of free-electron solids

2.1 Work out the free-electron value of the heat capacity constant γ for Cu, and compare with the experimental value of $97.3\,\mathrm{J\,m^{-3}\,K^{-2}}$.

2.2 If $E_x = \rho_{xx}J_x + \rho_{xy}J_y$ and $E_y = \rho_{yx}J_x + \rho_{yy}J_y$, show (by eliminating J_y) that $J_x = \sigma_{xx}E_x + \sigma_{xy}E_y$, and derive expressions for the conductivity components σ_{ij} in terms of the resistivity components ρ_{ij}. Show that if $\rho_{xx} = \rho_{yy}$ and $\rho_{yx} = -\rho_{xy}$, then $\sigma_{xx} = \sigma_{yy}$ and $\sigma_{yx} = -\sigma_{xy}$. Hence verify that the equations (2.21) are consistent with (2.23).

2.3 How big are τ and l for Cu at 300 K? How big is $\omega_c\tau$ if $B = 1$ T? What is then the angle between $\boldsymbol{J}$ and $\boldsymbol{E}$ if $\boldsymbol{J}$ is at right angles to $\boldsymbol{B}$? At low temperatures, the resistivity ρ may fall by a factor of 1000, in pure enough material. What effect would this have on the answers?

2.4 Verify that the integration of (2.25) yields (2.26).

2.5 For Cu, optical measurements give $\varepsilon_{\text{eff}} = -53\lambda^2$ for a free-space wavelength (measured in μm) between 1 μm and 10 μm (cf. Fig. 2.5). Use (2.34) to deduce a value for $\omega_p\varepsilon_r^{1/2}$, and compare this with the value predicted by (2.35).

2.6 Show that (2.39) is equivalent to (2.37) if $\boldsymbol{v}_d$ is independent of $\boldsymbol{v}$, but not otherwise. (Assume $\boldsymbol{v}_d = v_{x,d}, 0, 0$, and integrate (2.39) over v_x by parts.)

3. Crystal structures and the reciprocal lattice

3.1 Show that (3.1), with $\boldsymbol{a}_1, \boldsymbol{a}_2, \boldsymbol{a}_3$ given by (3.2) or (3.3), does indeed define every lattice point in a bcc or fcc lattice.

3.2 Suppose a hexagonal metal has $\sigma_{\parallel} = 2\sigma_{\perp}$. What is the largest possible angle between $\boldsymbol{J}$ and $\boldsymbol{E}$, and for what direction of $\boldsymbol{E}$ does it occur?

3.3 Show that the vectors (3.10) satisfy the conditions (3.8), and verify the result (3.11).

3.4 Using (3.10) and (3.2) or (3.3), work out a set of primitive reciprocal lattice vectors for the bcc and fcc real-space lattices. Your results will not at first sight look like fcc and bcc primitive vectors, but suitable linear combinations of primitive vectors can also be used as primitive vectors, and you should be able in this way to form the vectors $\boldsymbol{b}_1, \boldsymbol{b}_2, \boldsymbol{b}_3 = 2\pi(\boldsymbol{i}+\boldsymbol{k})/a,\ 2\pi(\boldsymbol{j}+\boldsymbol{k})/a,\ 2\pi(\boldsymbol{i}+\boldsymbol{j})/a$ for the bcc lattice and $4\pi\boldsymbol{i}/a$, $4\pi\boldsymbol{j}/a$, and $2\pi(\boldsymbol{i}+\boldsymbol{j}+\boldsymbol{k})/a$ for the fcc lattice.

4 Electrons in a periodic potential

4.1 Show that for $\boldsymbol{k} = \boldsymbol{k}_{\parallel}$, (4.19) can be written $(E - x^2)[E - (2-x)^2] = U^2$, where E, U, x are dimensionless variables related to $\varepsilon, V_1, k_{\parallel}$, and that if $x = 1 - \delta$, the solutions are $E = 1 + \delta^2 \pm \sqrt{(U^2 + 4\delta^2)}$. Plot $E(\delta)$ for $0 < \delta < 1$, for $U = 0$ (free electrons) and for $U = 0.1$, and confirm that the difference becomes small for $\delta \gtrsim U$.

4.2 Show that the contours of Fig. 4.3 are given by the equation $1 + \delta^2 \pm \sqrt{(U^2 + 4\delta^2)} + r^2 = \text{constant}$, where r is a dimensionless variable related to $k_{\perp}$.

4.3 We saw in Fig. 4.5 that the lowest band was completely filled in the cross-section $k_z = 0$. Is the *whole* band completely filled, or are there parts of the band, for $k_z \neq 0$, which are empty? If so, where, and roughly how big are they?

5 Electronic band structures

5.1 Use (4.9) to verify (5.6) and (5.7). (Note that the volume of the fcc BZ is just half the volume of the cube enclosing it, as for the bcc Wigner–Seitz cell in real space.) For what value of n does the free-electron sphere just touch the zone corner at W?

5.2 How many van Hove singularities in $g_n(\varepsilon)$ would you expect sc, bcc and fcc crystals to show?

6 The potential $V(\boldsymbol{r})$; many-body effects

6.1 A determinant changes sign if two rows, or two columns, are interchanged, and must therefore vanish if two rows or two columns are identical. Applying these results to (6.9), show that Ψ satisfies the antisymmetry condition (6.6), (6.7), and that Ψ vanishes if any two functions ψ_i and ψ_j are identical; that is, if two electrons occupy the same state with the same spin.

For two parallel-spin electrons, (6.9) reduces to $\Psi = \psi_1(\boldsymbol{r}_1)\psi_2(\boldsymbol{r}_2) - \psi_1(\boldsymbol{r}_2)\psi_2(\boldsymbol{r}_1)$. Show that if $\psi = \mathrm{e}^{i\boldsymbol{k}_1\cdot\boldsymbol{r}}$ and $\psi_2 = \mathrm{e}^{i\boldsymbol{k}_2\cdot\boldsymbol{r}}$, Ψ falls smoothly to zero as $\boldsymbol{r}_1$ approaches $\boldsymbol{r}_2$, and thus shows an 'exchange hole'.

7 The dynamics of Bloch electrons

7.1 Suppose that a pure Cu sample at 4.2 K has $\rho = 20 \times 10^{-12}$ ohm m, about 1/1000 of the room-temperature value, and suppose that the need to keep it cold limits the power dissipation to 1 W cm^{-3}. What current density J and what field E does this correspond to? On the free-electron model, what is the drift velocity v_d and the corresponding displacement k_d of the FS? How big is k_d compared with the dimensions of the BZ?

7.2 Show that for free electrons, $m_\mathrm{c}^* = m$, where m_c^* is given by (7.7).

7.3 Show that in (7.12), $\oint \hbar\boldsymbol{k}\cdot\mathrm{d}\boldsymbol{r} = 2e\boldsymbol{B}\cdot\mathscr{A}_\mathrm{r}$, and $\oint e\boldsymbol{A}\cdot\mathrm{d}\boldsymbol{r} = e\boldsymbol{B}\cdot\mathscr{A}_\mathrm{r}$ where $\mathscr{A}_\mathrm{r}$ is the area enclosed by the projection of the orbit on the plane normal to $\boldsymbol{B}$. (Use (7.8) to write $\boldsymbol{k}$ in terms of $\boldsymbol{r}$, and use Stokes' theorem to express $\oint \boldsymbol{A}\cdot\mathrm{d}\boldsymbol{r}$ in terms of curl $\boldsymbol{A} = \boldsymbol{B}$. Note that the vectors $\mathscr{A}_\mathrm{r}$ and $\boldsymbol{B}$ are parallel.)

7.4 Verify that for free electrons, (7.14) leads to (7.10).

7.5 Treating Cu as a free-electron metal, use the data of problem 1.2 to work out the area $\mathscr{A}$ of the largest $\boldsymbol{k}$-space orbit for electrons at the FS. The largest fields B readily attainable with iron-cored magnets are about 2 T, and with superconducting solenoids about 10 T. Work out the quantum number n of (7.14) for this value of $\mathscr{A}$ and these values of B.

7.6 Show that for a free-electron metal the occupied length Δk_z of a Landau tube of csa $\mathscr{A}_n$ is given by $\Delta k_z = 2\sqrt{[(\mathscr{A}_e - \mathscr{A}_n)/\pi]}$, where $\mathscr{A}_e$ is the extremal csa of the FS. How does Δk_z vary with B?

7.7 Show that if the csa of the FS near $k_z = 0$ is written in the form $\mathscr{A} = \mathscr{A}_e + \frac{1}{2}\mathscr{A}_e'' k_z^2$ (where $\mathscr{A}_e''$ is negative if $\mathscr{A}_e$ is a maximum area), then $\Delta k_z = \sqrt{[8(\mathscr{A}_e - \mathscr{A}_n)/|\mathscr{A}_e''|]}$. What happens if $\mathscr{A}_e$ is a minimum area?

7.8 Calculate the value of $e^{-\alpha T}$ in (7.18a) when $m_c^* = m$ (the free-electron mass), $B = 10\,T$ and $T = 2\,K$. Use the results of problem 2.3 to calculate $e^{-\alpha T_D}$ for Cu when $B = 10\,T$ and $\rho = \rho_{RT}/1000$.

8 Collisions

8.1 Use (7.2) to show that if $|V_{kk'}|$ in (8.5) is assumed to be independent of $\boldsymbol{k}'$, the total scattering rate (for $n_i = 1$) becomes

$$\int P_{kk'} d^3\boldsymbol{k}' = (2\pi/\hbar)|V_{kk'}|^2 g_s(\varepsilon_k)$$

where $g_s(\varepsilon_k) = \frac{1}{2}g(\varepsilon_k)$ is the density of states for spin-up or spin-down electrons. This is the form in which Fermi's 'Golden Rule' is often quoted.

8.2 Verify that for a spherically symmetric potential, (8.3) reduces to (8.6). Show that inserting (8.7) or (8.8) in (8.6) yields $V_{kk'}/V_{kk'}(0) =$ (a) $1/(1 + x_1^2)$, (b) $3(\sin x - x\cos x)/x^3$ where $x_1 = k_s r_1$, $x = k_s r_0$ and $V_{kk'}(0)$ is given by (8.11).

8.3 Show that when proper account is taken of the V_r factors, both (8.4) and (8.15) give $P_{kk'}\delta^3\boldsymbol{k}'$ the dimensions of $(\text{time})^{-1}$. (Note that $\delta(\varepsilon_{k'} - \varepsilon_k)$ has dimensions of $(\text{energy})^{-1}$, and that whereas n_q is dimensionless, n_i has dimensions $(\text{length})^{-3}$.)

8.4 The proportion of its time that an electron spends in free paths of duration between t and $t + \delta t$ is $P(t) = t\,p(t)\delta t/\int_0^\infty t\,p(t)dt$ (cf. (8.22)).

An electron picked at random will therefore be on average in a free path of duration $\int_0^\infty t\,P(t)\,dt$. Verify that this is equal to 2τ.

9 Electrical conductivity of metals

9.1 As a slight generalization of the free-electron model, consider a metal for which $\varepsilon = \hbar^2k^2/2m^*$ (where m^* may differ from the free electron mass m), so that $\hbar\boldsymbol{k} = m^*\boldsymbol{v}$. Show that if $\boldsymbol{L}_{\boldsymbol{k},\parallel} = \boldsymbol{v}\tau$, (9.11) reduces at once to $\sigma = ne^2\tau/m^*$.

9.2 For $T > \theta_{\mathrm{D}}$, we can write $n_{\boldsymbol{q}}(q^2/\omega_{\boldsymbol{q}}v_{\boldsymbol{k}'}) = kTq^2/\hbar\omega_{\boldsymbol{q}}^2v_{\boldsymbol{k}'} = kT/\hbar v_{\mathrm{s}}^2 v_{\mathrm{F}}$, if we use the approximation $\omega_{\boldsymbol{q}} = v_{\mathrm{s}}q$, and assume an NFE model so that $v_{\boldsymbol{k}'} = v_{\mathrm{F}}$ for all $\boldsymbol{k}'$. With these approximations, this term becomes independent of $\boldsymbol{k}'$ and can be taken outside the integral in (9.15), which then reduces to $\int \mathrm{d}S_{\boldsymbol{k}'}\,(1 - \cos\theta)$, $= 4\pi k_{\mathrm{F}}^2$ on the NFE model. Putting $V_0 = 2\varepsilon_{\mathrm{F}}/3$, (9.15) thus reduces to

$$1/\tau^{\mathrm{e}} = (2\varepsilon_{\mathrm{F}}k_{\mathrm{F}}/3\hbar v_{\mathrm{s}})^2\,kT/\pi D v_{\mathrm{F}}$$

where $D = MN$ is the density of the metal. Use this approximation to estimate τ^{e} for Cu at 300 K, with the values of ε_{F}, k_{F} and v_{F} from problem 1.5. Take $v_{\mathrm{s}} = 3600\ \mathrm{m\,s^{-1}}$ and $D = 8.9 \times 10^3\ \mathrm{kg\,m^{-3}}$. Compare your result with the value of τ found in problem 2.3.

9.3 Show that if the ratio $\tau_{\boldsymbol{k},r}^{\mathrm{e}}/\tau_{\boldsymbol{k},t}^{\mathrm{e}} = \alpha(T)$ is independent of $\boldsymbol{k}$, and if (9.17) holds, then (9.13) leads to the result $\rho(= 1/\sigma) = \rho_r + \rho_t(T)$, where ρ_r is determined by defect scattering and $\rho_t\,(= \alpha\rho_T)$ by phonon scattering.

9.4 To see what happens if the ratio α in problem 9.3 varies with $\boldsymbol{k}$, suppose that $\alpha = \alpha_0(1 + x)$ over half the FS, and $\alpha_0(1 - x)$ over the other half. The metal then 'consists effectively of two metals in parallel, for which $\rho = 2[\rho_r + \rho_t(1 + x)]$ and $2[\rho_r + \rho_t(1 - x)]$ respectively (where $\rho_t = \alpha_0\rho_r$). Show that the resulting total resistivity will then be $\rho_r + \rho_T + \rho_r\rho_t x^2/(\rho_r + \rho_t)$, where $\rho_T = \rho_t(1 - x^2)$. Thus if $x = 0.1$, so that $\rho_t(1 + x)$ and $\rho_t(1 - x)$ differ by 20%, the departure from Matthiessen's rule (i.e. from $\rho = \rho_r + \rho_T$) is at most 0.25%, when $\rho_r = \rho_t$.

10 Metals in a temperature gradient

10.1 Derive (10.7). (Hint: replace the lower limit by $-L$, with L large, integrate by parts, and use (1.15) with the lower limits again replaced by $-L$.)

10.2 Verify that (10.20) and (10.21) follow from (10.18) and (10.19), and that (10.27) follows from (10.20), (10.21) and (10.25).

11 Magnetoresistance and Hall effect

11.1 Starting from $J_i = \sum_j \sigma_{ij} E_j$ and $E_i = \sum_j \rho_{ij} J_j$, verify the results (11.3).

11.2 Verify that for free electrons (11.8) reduces to (11.9), with $\sigma_0 = ne^2\tau/m$.

11.3 Verify the two-band expressions (11.15) and (11.16), and verify that the explicit field-dependence of ρ_T and R_T vanishes if $R_1/\rho_1 = R_2/\rho_2$. This is the condition for the two Hall angles ϕ_1 and ϕ_2 to be equal, since $\tan\phi = RB/\rho$. Show that $\Delta\rho_T/\rho_T(0) = [\rho_T(B) - \rho_T(0)]/\rho_T(0)$ can be written in the form

$$\frac{\Delta\rho_T}{\rho_T(0)} = \frac{(\rho_1 R_2 - \rho_2 R_1)^2 B^2}{\rho_1\rho_2[(\rho_1+\rho_2)^2 + (R_1+R_2)^2 B^2]}$$

and is thus always positive.

12 Radio-frequency, optical and other properties

12.1 Starting from (12.9), verify that for a free-electron metal $\sigma_{\text{eff}} = 3\beta\delta_{\text{eff}}\sigma_0/2l$, where $\sigma_0 = ne^2 l/mv_F$. By combining (12.9) with (2.33), show that $\delta_{\text{eff}} = (\mu_0\omega\beta s)^{-1/3}$, and that Z is given by (12.10).

For the free-electron metal of problem 1.5, work out δ_{eff} and R (where $Z = R + iX$) for frequencies $\omega/2\pi$ of 1, 10^2 and 10^4 MHz, assuming $\delta_{\text{eff}} \ll l$ and taking $\beta = 1.5$. How large must l be if the condition $\delta_{\text{eff}} \ll l$ is to be satisfied at these frequencies?

12.2 Use (12.2) to show that for a spatially uniform field $\boldsymbol{E} = E_x e^{i\omega t}$, $\Delta\varepsilon_{\boldsymbol{k}} = -eE_x(v_x\tau)_{\boldsymbol{k}}/(1 + i\omega\tau_{\boldsymbol{k}})$, and hence show that for $\omega\tau_{\boldsymbol{k}} \gg 1$, $\bar{\sigma} = \frac{1}{3}(\sigma_{xx} + \sigma_{yy} + \sigma_{zz})$ is given by (12.12).

13 Carriers in semiconductors

13.1 Use (13.9) to find the values of n_e and n_h in intrinsic Si at 200, 300 and 400 K, given that $\varepsilon_g = 1.11$ eV. Take the six electron ellipsoids to have effective mass components of $0.97m$, $0.19m$ and $0.19m$, and take the heavy and light holes to have effective masses $m^*_{h,1}$ and $m^*_{h,2}$ of $0.50m$ and $0.16m$, where m is the free electron mass.

How do these values of n_e and n_h compare with the value of n_e

for a typical metal such as Cu? How much would n_e and n_h be altered if we included in N_h an allowance for the third hole band (13.12), with $\varepsilon_3 = 0.035$ eV and $m^*_{h,3} = 0.25m$?

13.2 Show that $[1 + \frac{1}{2}\exp(\varepsilon - \varepsilon_F)/kT]^{-1} \approx f_0(\varepsilon - 0.7kT)$, where $f_0(\varepsilon) = [1 + \exp(\varepsilon - \varepsilon_F)/kT]^{-1}$. Show that if ε_F lies in the gap between the donor and acceptor levels, far enough from either for f_0 and $1 - f_0$ to be replaced where appropriate by their exponential tails, $N_{d,e}N_{a,h} = N_d N_a \exp - (\varepsilon_g - \varepsilon_d - \varepsilon_a)/kT$. Verify (13.15).

14 Transport properties of semiconductors

14.1 Find the value of $\alpha = 2kr_1$ in (14.1) (where $\varepsilon_k = \hbar^2 k^2/2m^*$ and r_1 is given by (8.10)), if $m^* = 0.1m$, $\varepsilon_k = kT$, with $T = 290$ K, $\varepsilon_r = 12$ and $n = 10^{22}$ carriers m^{-3}. Again assuming $\varepsilon_k = kT$ with $T = 290$ K, use this value of α to find τ^e_k from (14.2), if $N_I = 10^{22}\ m^{-3}$. Find v_k and hence $l_k = \tau^e_k v_k$. Finally, find the value of $\alpha_1 = 4\pi\varepsilon_r\varepsilon_0\varepsilon_k/e^2 N_I^{1/3}$.

14.2 Show that if the energy surfaces have the ellipsoidal form (13.2), if $f_0(\varepsilon) = C\ e^{-\varepsilon/kT}$, and if we put $\tau(\varepsilon) = \tau_0 u^\nu$ where $u = \varepsilon/kT$, (14.5) reduces to

$$\sigma_{xx} = A(e^2\tau_0/m^*_x)\int_0^\infty u^{3/2+\nu}\,e^{-u}\,du$$

where $A = (8m^*_x m^*_y m^*_z)^{1/2} C(kT)^{3/2}/3\pi^2\hbar^3$.

Show also that the number of electrons, $n_e = \int d^3\boldsymbol{k} f_0/4\pi^3$, can be written

$$n_e = \tfrac{3}{2}A\int_0^\infty u^{1/2}e^{-u}\,du = A\int_0^\infty u^{3/2}e^{-u}\,du$$

where the second form follows on integration by parts. Hence confirm (14.6) and (14.7).

[The *gamma function* $\Gamma(n+1) = \int_0^\infty u^n\ e^{-u}\,du$ can be found in mathematical tables. Integration by parts shows that $\Gamma(n+1) = n\Gamma(n)$, so that if n is an integer, $\Gamma(n+1) = n(n-1)(n-2)\ldots = n!$ For $n = \frac{1}{2}$, $\Gamma(\frac{1}{2}) = 2\int_0^\infty e^{-x^2}\,dx = \sqrt{\pi}$ (putting $u = x^2$).]

14.3 Show that for the ellipsoidal surface (13.2), $m^*_c = (m^*_y m^*_z)^{1/2}$ if $B = B_x$. How could electron and hole surfaces be distinguished by cyclotron resonance experiments?

14.4 To see the effect of orbit quantization in a two-dimensional metal when the number of electrons per unit area n_e is fixed, consider a free-electron-like metal in which $\varepsilon = \hbar^2k^2/2m^*$ when the sample is large (and when $B = 0$). Show that for a sample of thickness Δz, in which $\psi = 0$ at $z = 0$ and $z = \Delta z$, the lowest states have energy $\varepsilon = \hbar^2(k_x^2 + k_y^2)/2m^* + \varepsilon_z$ when $B = 0$, where $\varepsilon_z = \pi^2\hbar^2/2m^*\Delta z^2$. In a field B_z, the allowed values of ε are thus $\varepsilon_r = (r - \frac{1}{2})\hbar\omega_c + \varepsilon_z$, if the lowest Landau level is labelled $r = 1$, and if we *neglect* the effect of spin splitting. If the degeneracy of each level is $2eB/h$ (again neglecting spin splitting), show that as B increases from $n_eh/2er$ to $n_eh/2e(r-1)$, ε_F increases from $\varepsilon_{F,0} - \frac{1}{2}\hbar\omega_c$ to $\varepsilon_{F,0} + \frac{1}{2}\hbar\omega_c$ (where $\varepsilon_{F,0}$ is the value of ε_F when $\boldsymbol{B} = 0$), and then falls back abruptly to $\varepsilon_{F,0} - \hbar\omega_c$ as the rth level empties. (cf. Fig. 7.5: the spacing between the levels increases linearly with B.)

14.5 Problem 14.4 shows that in a two-dimensional sample with fixed n_e, ε_F must fluctuate as B_z varies. In terms of Fig. 14.4b, this means that the distance $\Delta\varepsilon$ from the bottom of the band to the Fermi level must fluctuate. In fact $\Delta\varepsilon$ is determined by the applied electric field $E_{z,0}$ at the surface of the semiconductor and by the distribution of charge in the surface layer. Suppose, as a rough approximation, that the charge n_ee per unit area is uniformly distributed within a layer of thickness Δz and is zero outside the layer. Show that on this approximation $\Delta\varepsilon = \frac{1}{2}eE_{z,0}\Delta z - (\varepsilon_{c,\infty} - \varepsilon_F)$, where $\varepsilon_{c,\infty}$ is the value of ε_c at large z.

The necessary fluctuations in $\Delta\varepsilon$ are thus produced by fluctuations in the layer thickness Δz.

14.6 To see the effect of localized states, suppose for simplicity that a fraction α of the surface area in Fig. 14.7 is covered by localized states, and a fraction $(1 - \alpha)$ by extended states; and suppose that the extended states have the unbroadened energies ε_r of problem 14.4, with degeneracy $2eB/h$ per unit area (again neglecting spin splitting), while the localized states form a continuum, so that the number per unit area up to energy ε is $m^*(\varepsilon - \varepsilon_z)/\pi\hbar^2$. Show that when $n_eh/2eB$ lies in the range $r \pm \frac{1}{2}\alpha$, ε_F will lie *between* the extended-state levels ε_r and ε_{r+1} (so that $\rho_{xy} = \rho_0/r$, independent of B).

RELEVANT FUNDAMENTAL CONSTANTS

Speed of light	$c = 3.00 \times 10^{8}\ \mathrm{m\,s^{-1}}$
Charge on electron $-e$, where	$e = 1.60 \times 10^{-19}\ \mathrm{C}$
Planck's constant	$h = 6.63 \times 10^{-34}\ \mathrm{J\,s}$
$h/2\pi$	$\hbar = 1.05 \times 10^{-34}\ \mathrm{J\,s}$
Boltzmann's constant	$k = 1.38 \times 10^{-23}\ \mathrm{J\,K^{-1}}$
Mass of electron	$m = 9.11 \times 10^{-31}\ \mathrm{kg}$
Avogadro's number	$N_A = 6.02 \times 10^{23}\ \mathrm{mol^{-1}}$
Permittivity of free space	$\varepsilon_0 = 8.85 \times 10^{-12}\ \mathrm{C\,V^{-1}\,m^{-1}}$

Answers to problems

1 The free-electron model

1.1 $1.17 \times 10^5\,\mathrm{m\,s^{-1}}$

1.2 $n = 8.4 \times 10^{28}\,\mathrm{m^{-3}}$; hence $l = 2.9\,\mathrm{nm}$ or $5.7\,\mathrm{nm}$

1.4 Write $1/(e^x + 1) = e^{-x} - e^{-2x} + e^{-3x} \ldots$

1.5 $\varepsilon_{F,0} = 1.115 \times 10^{-18}\,\mathrm{J} = 6.97\,\mathrm{eV}$; $k_{F,0} = 1.356 \times 10^{10}\,\mathrm{m^{-1}}$; $v_{F,0} = 1.56 \times 10^6\,\mathrm{m\,s^{-1}}$

2 Properties of free-electron solids

2.1 $71.0\,\mathrm{J\,m^{-3}\,K^{-2}}$

2.3 At 300 K, $\tau = 2.45 \times 10^{-14}\,\mathrm{s}$; $l = 38.3\,\mathrm{nm}$; $\omega_c = 1.76 \times 10^{11}\,\mathrm{s^{-1}}$; $\omega_c\tau = 4.3 \times 10^{-3}$; $\phi = \tan^{-1}(\omega_c\tau) = 0.25°$. If ρ falls by 1000, $\tau = 2.45 \times 10^{-11}\,\mathrm{s}$; $l = 38.3\,\mu\mathrm{m}$; $\omega_c\tau = 4.3$; $\phi = 77°$

2.5 $1.37 \times 10^{16}\,\mathrm{Hz}$; $1.64 \times 10^{16}\,\mathrm{Hz}$

3 Crystal structures and the reciprocal lattice

3.2 19.5°; 54.7° to c axis

5 Electronic band structures

5.1 2.93

7 The dynamics of Bloch electrons

7.1 $J = 2.23 \times 10^4\,\mathrm{A\,cm^{-2}}$; $E = 4.46\,\mathrm{mV\,m^{-1}}$; $v_d = 1.65 \times 10^{-2}\,\mathrm{m\,s^{-1}}$; $k_d = 143\,\mathrm{m^{-1}} \approx 10^{-8} \times$ size of BZ

7.5 $k_F = 1.356 \times 10^{10}\,\mathrm{m}^{-1}$, so $\mathscr{A} = 5.78 \times 10^{20}\,\mathrm{m}^{-2}$. $n \approx 3 \times 10^4$; 6×10^3

7.8 0.052; 0.93

9 Electrical conductivity of metals

9.2 $1.48 \times 10^{-14}\,\mathrm{s}$

12 Radio-frequency, optical and other properties

12.1 $\delta_{\mathrm{eff}} = 4.2, 0.91, 0.195\,\mu\mathrm{m}$; $R = 0.017, 0.36, 7.7\,\mathrm{m}\Omega$; $l \gtrsim 40, 9, 2\,\mu\mathrm{m}$

13 Carriers in semiconductors

13.1 9.92×10^{10}, 8.29×10^{15}, $2.72 \times 10^{18}\,\mathrm{m}^{-3}$. In Cu, $n_e = 8.43 \times 10^{28}\,\mathrm{m}^{-3}$. Including the effect of the third hole band increases n_e, n_h by 2% at 200 K and by 5% at 400 K

14 Transport properties of semiconductors

14.1 $\alpha = 6.9$; $\tau^e = 1.1 \times 10^{-12}\,\mathrm{s}$; $l = 1.0\,\mu\mathrm{m}$; $\alpha_1 = 10.0$

14.3 By using a circularly polarized field $\boldsymbol{E}$, and reversing the direction of $\boldsymbol{B}$.

Further reading

There are many texts on solid-state physics which include something on the topics covered in this book, and the following list is by no means exhaustive.

Kittel, C. (1986). *Introduction to Solid State Physics*, 6th edn, Wiley, Chichester.
When the first edition of this book was published in 1953, it at once established itself as the standard text in the field, and remained so for many years. Still useful for reference.

Blakemore, J.S. (1985). *Solid State Physics*, 2nd edn, Cambridge University Press, Cambridge.
Covers much the same ground as Kittel, but fuller on semiconductors. A good alternative to Kittel, and far cheaper.

Dugdale, J.S. (1977). *Electrical Properties of Metals and Alloys.* Edward Arnold, Sevenoaks.
A detailed treatment of transport properties.

Elliott, R.J. and Gibson, A.F. (1974). *Introduction to Solid State Physics and its Applications.* Macmillan, London.
Rather densely packed with information. Good on optical properties and applications.

McClintock, P.V.E., Meredith, D.J. and Wigmore, J.K. (1984). *Matter at Low Temperatures.* Blackie & Son, Edinburgh.
A very clear, concise and readable treatment of a wide range of topics.

Ashcroft, N.W. and Mermin, N.D. (1976). *Solid State Physics.* Holt, Rinehart and Winston, New York.
Excellent as a more advanced text. A large book, but very clear and readable, and explains difficulties which others ignore.

Ziman, J.M. (1972). *Principles of the Theory of Solids*, 2nd edn. Cambridge University Press, Cambridge.

A text for graduate students. Covers a lot of ground well but briefly.

Pippard, A.B. (1989). *Magnetoresistance in Metals*, Cambridge University Press, Cambridge.

Shoenberg, D. (1984). *Magnetic Oscillations in Metals*, Cambridge University Press, Cambridge.

These two research monographs deal in depth with the topics of Chapter 11 and Chapter 7 (Section 7.4), respectively.

Mott, N.F. (1987). *Conduction in Non-Crystalline Materials*, Oxford University Press, Oxford.

A very individual treatment of some current research topics, by one of the earliest leaders in the field. Over fifty years earlier, in 1936, Mott and Jones had written one of the first classics in the field, *The Theory of the Properties of Metals and Alloys*, Oxford University Press.

Index